SCIENTIST AND CATHOLIC

An Essay on Pierre Duhem

By the same author

Les tendances nouvelles de l'ecclésiologie

The Relevance of Physics

Brain, Mind and Computers
(Lecomte du Nouy Prize, 1970)

The Paradox of Olbers' Paradox

The Milky Way: An Elusive Road for Science

*Science and Creation: From Eternal Cycles
to an Oscillating Universe*

*Planets and Planetarians: A History of Theories
of the Origin of Planetary Systems*

The Road of Science and the Ways to God
(Gifford Lectures: University of Edinburgh, 1975 and 1976)

The Origin of Science and the Science of its Origin
(Fremantle Lectures, Oxford, 1977)

*And on This Rock: The Witness of One Land
and Two Covenants*

Cosmos and Creator

Angels, Apes and Men

Uneasy Genius: The Life and Work of Pierre Duhem

Chesterton: A Seer of Science

The Keys of the Kingdom: A Tool's Witness to Truth

Lord Gifford and His Lectures: A Centenary Retrospect

Chance or Reality and Other Essays

The Physicist as Artist: The Landscapes of Pierre Duhem

(continued on p. 279)

SCIENTIST and CATHOLIC

An Essay on Pierre Duhem

STANLEY L. JAKI

Christendom Press

Copyright © 1991 Stanley L. Jaki

Jaki, Stanley L. (1924–)
    Scientist and Catholic: An Essay on Pierre Duhem

    1. Life and thought of P. Duhem 2. Science and catholic faith

ISBN 0-931888-44-1

Published in the United States by
Christendom Press
Christendom College
Front Royal, VA 22630

Contents

Introduction

This essay owes its immediate origin to the concern of Monsieur Jacques Vauthier, Head of the Department of Mathematics of the Sorbonne (Université de Paris IV) and Professor at the Institut Henri Poincaré, Paris. As the editor of the series "Hommes de foi et de science" (Beauchesne, Paris) he pressed me to contribute to that series a volume on Pierre Duhem.* He did so in connection with a conference which I gave, in early December 1989, on Duhem the landscape artist at Collège Stanislas which counts Duhem among its most illustrious alumni and where Professor Vauthier serves as President of the Alumni Association. The essay was not to emulate in scope, length, and documentation an earlier work of mine, *Uneasy Genius: The Life and Work of Pierre Duhem*, which Nijhoff Publishers brought out in a paperback edition in 1987, following its first publication in 1984. To be sure, Duhem the Catholic stands out in that book which is, however, above all about the physicist he always wanted to be. Here a direct emphasis is laid on Duhem the truly devout and thoroughly convinced Catholic and with the help of epistolary and other evidences that could not be available at the writing of *Uneasy Genius*.

Had Duhem been but an exemplary practising Catholic, an account of his life and work focused on that fact would need

7

no further justification. We live in a secularist age which is not willing to learn from the colossal debacle of institutionalized Marxist slogans about religion as a mere opiate of the people. By and large, scientists in the Western world fail to protest against those colleagues of theirs who, blessed with literary and performing talents, keep preaching a now two-hundred-year-old message of secularist Western culture. The message, first formulated by the gurus of the Enlightenment, consists in the claim that science is the only reliable savior of mankind and that for science to be born Christianity, or the religion most explicitly steeped in belief in a most extraordinary Savior, first must be discredited.

That religion, including Christianity in general and Catholicism in particular, can only be tolerated as a subjective option, is the implicit message of pontificating scientists, all too ready to perform before the batteries of television cameras. The option they allow to that religion is a lame licence to operate as an opiate which, so they hope, proper exposure to science will sufficiently prevent from doing too much harm.

Against such message and claim it is always effective to fall back on Diogenes' unsurpassable method of refutation. He was not the first to be presented with the sophisticated fallacy (then as always coated in specious references to the method of science) aimed at securing the rule of universal scepticism. The fallacy was meant to prove that it was impossible to traverse any distance, however small, because any such distance consists of an infinite number of parts which it takes an infinite time to traverse. Apparently, Diogenes was the first to dispose of that hollow argument by walking from one end of a room to the other. His commonsense act became remembered in terms of a proverbial phrase, *solvitur ambulando*. The very life of any scientist who was also a devout Christian is such a perennial factual disproof of that impertinent secularist message. Such a disproof can particularly well serve those who cannot handle effectively the arguments, philosophical and historical, that invariably enter the scene whenever the question arises about the alleged conflict between science and religion.

Duhem, the scientist and Catholic, can be utilized in a far deeper sense. For Duhem was unique among modern scientists with his penetrating insights into the method of the exact sciences, and in particular of physics, both on the conceptual level and along the vast and broad front of its use in history. In fact he did, what historians and historians of science were supposed to have done long ago: He discovered the true origins of Newtonian physics. That those origins are steeped in a culture, the Middle Ages, which for many is still the classic embodiment of obscurantism, could have but served as *lèse majesté*. But as if insult were to be added to injury, Duhem also spelled out the fact, with a vast and most original historical research that those origins are intimately connected with Catholic dogmas, such as the creation out of nothing and creation in time.

Catholics, because it is for them above all that this essay is written, should now pause. They are, of course, utterly mistaken if they expect Catholic facts to prevail in secularist consciousness. Duhem or not, the academic milieu, to say nothing of its journalistic overspill, will continue in the merry belief that science had forever disposed of the possibility and fact of Revelation, especially as given in Jesus Christ, the only LORD. It is on that merry belief of theirs that rests the basic dogma of secularism, namely, that man is his own master, accountable to no one on this earth, let alone above it.

Against such a milieu, which threatens him with continual malaise and periodic suffocation, the Catholic needs a solid antidote. It cannot consist of bad poetizing in good prose about a "divine milieu," to say nothing of a "cosmic Christ" who cannot be a Redeemer and Savior because, in the alleged absence of original sin, nothing serious remains for Him to do. The solid antidote can only consist in rigorous thinking and mastery of incontestable facts. It is these that Duhem provided in the teeth of an at times ferocious opposition and against extraordinary odds. He coped with them because he lived his Catholic faith in a measure that was far beyond ordinary.

It was Duhem's deep conviction that Divine Providence rules everything (he would have had only contempt for the glorification of chance in terms of a widespread misinterpretation of quantum mechanics). It may not therefore be presumptious to think that the same Providence determined this Introduction to be written around the time, September 28, 1990, the hundredth anniversary of Duhem's wedding to Marie Adèle Chayet in Saint Sulpice, Paris. The excruciating blow, which he suffered when his wife passed away in her second and unsuccessful childbirth after less than two years of a most happy marriage, Duhem bore for the rest of his life with deep resignation in God's inscrutable though providential will. The pain he had always felt on account of having lost his beloved wife may have heavily contributed, in addition to unrelenting hard work and intestinal rheumatism, to the gradual weakening of his heart. But long before he was felled by a fatal heart attack on September 14, 1916, he must have been able to scrutinize the ways of divine Providence. Had he not been destined to live a solitary life, he could not have made the search for scientific truth (catholic as well as Catholic) the sole purpose of his heroic life.

S. L. J.

*Since the idea of this book originated in France, and in particular in Paris which Duhem always held to be his real home, it was agreed that the English-language edition should follow by six months the publication of the French version, *Pierre Duhem: Homme de foi et de science* (Paris: Beauchesne).

1

The Setting

In speaking about Pierre Duhem as a man of science and man of faith, it is impossible not to think first of what may be his most memorable statement:

> Of course, I believe with all my soul in the truths that God has revealed to us and that He has taught us through His Church; I have never concealed my faith, and that He in whom I hold it will keep me from ever being ashamed of it, I hope from the bottom of my heart.

Such were the words Duhem put almost at the very start of his long and famous essay, "Physics of a Believer."[1] At its end Duhem registered the place, Peyreleau, where he wrote it, and the date, 9 September 1905, when he completed it. A year later, almost to the day, he was back in that quaint village and made a magnificent drawing of it as it is overshadowed by the almost vertical mountainside at the confluence of the Tarn and of the Jonte.[2]

The famous gorges of those two rivers he chose repeatedly for his September hikes, his only form of recreation. Hiking kept him only from his writing desk, not from his constant reflections — scientific and philosophical. Duhem, who com-

posed his writings in his head, found in walking a most
effective help for his mind to find answers to many a problem.
Such was the observation made about him by Edouard Jordan,
professor of medieval history at the Collège de France and a
friend of his from their days in the Collège Stanislas. Jordan
spoke on the basis of close range observation. He spent a year
(1893-94) with Duhem at the University of Rennes and was,
on several occasions, his companion through the Cévennes.

The "Physics of a Believer" has indeed that vibrating
character which is revealed by writings full of ideas that their
authors first recite with marked animation to themselves before
putting them on paper in a single effort. As one walks along
roads free of cars, posters, and vendors — and on roads that
around Peyreleau were a century ago often free even of horse-
drawn carriages and oxcarts — one can naturally engage in a
mental writing if the topic is very much on one's mind. This
was certainly the case with the message of "Physics of a
Believer." Duhem meant it to be an answer to an article on his
philosophy of science which its author, Abel Rey, had just
published in the Revue de métaphysique et de morale,[3] easily the
foremost of such periodicals in France at that time.

A young agrégé of the Sorbonne, Rey was in the process of
presenting the views of all the major philosophers of science
around the turn of the century. He had in mind much more
than a purely academic exercise. In fact he meant to write an
apologetic of materialist rationalism. The eventual proof of this
was a book he published twenty or so years later, Le retour
éternel et la philosophie de la physique. There he tried to restore
intellectual respectability to the quintessence of ancient paganism, the idea of an eternal universe going through endless cycles
with no beginning and end.[4]

Rey's strategy was rather simple, if not simplistic. Yet his
strategy seemed to have a built-in credibility in these modern
times when anything, however illusory or arbitrary, can
command attractiveness provided it is wrapped in copious

references to science. One wonders whether Rey, who by then shifted from the philosophy of science to its history, had ever perceived a supreme irony. His lengthy encomium of eternal returns was published in that very year, 1927, when the abbé Lemaître derived the rate of the expansion of the universe and provided, on a cosmic scale, a powerful support to the irreversibility of all physical processes, embodied in the law of entropy. Rey did his best to take the sting out of that law, by then firmly established for over half a century.

To Rey's credit, he did not leave his readers in the dark about his principle guiding light: belief in the eternity of the universe. His book came to a close with a quotation from Spinoza's *Ethics*: "Sentimus et experimur nos aeternos esse." Whether such an experience was more than a mere sentiment in order to qualify as a proper object of rational debate, Spinoza failed to discuss. Nor did Rey enter into the problems posed by Spinoza's mystical identification with the universe or Nature, writ large, a Nature raised to a divine pedestal. Such a universe was creative in the sense given to that word by Bergson, an unabashed pantheist when he wrote the *Evolution créatrice*. In fact, it was his reference to Bergson's creative universe that made Rey invoke that supreme pontiff of pantheism, Spinoza.

Pantheism was a half-acknowledged message of Whitehead's process philosophy and more recently, though in a covert way, of Popper's open universe. John Henry Newman proved himself a prophet, when in Tract Eighty-Five (September 1838) he spoke of pantheism as "the great deceit which awaits the age to come." Once more Newman was true to himself as he graphically portrayed the manner in which a patently un-Christian idea would present itself as the unfolding of the very best in Christianity. In only one respect did Newman fall short of his usually very high performance. He listed beauty, imagination, and philosophy, in addition to licence given to unrestrained speculation and self-indulgence, as so many tenets of a new Christianity liberated from its old self.

He failed to mention science as another pretext for choosing pantheism. So much about the deeper bearings of Rey's dressing up the idea of an eternal universe in paraphernalia with scientific glitter.

One wonders what Rey's feelings were when, following Bergson's funeral in 1914, news spread all across Paris and far beyond that Bergson had abandoned pantheism. It was at Bergson's specific request that at his funeral the Our Father was recited by Fr. Sertillanges. The illustrious Dominican would not have done so had he not learned of an all-important change in Bergson's thinking. The famed author of the *Evolution créatrice* came to acknowledge a personal transcendental Creator as the sole explanation of an evolution which is a move forward and not a moving around in a circle. The idea of such a Creator is certainly a pivotal part of the Christian Catholic creed which Bergson considered embracing formally through baptism. If he had not done so it was only because of his apprehension that his step would be taken by many as a cheap means of escape from the nets the Nazis were spreading around him and other Jews.

Rey was not, of course, unaware of a close identity between an eternal universe and a moving around in a circle. That he saw the idea of eternal recurrence in all its ramifications is clear from the last section in the last chapter of his book, a chapter dealing with the *reality* of eternal returns. It was replete with quotations from the works of Nietzsche, the self-anointed prophet of eternal returns and chief salesman of ancient paganism in his times. Well-advised readers of Nietzsche could easily suspect his utter incompetence in physics. It could be otherwise even with scientifically trained readers of Rey's book. Rey could easily appear to them utterly competent in his review of the scientific pros and cons concerning a universe which had to be eternal according to Rey's innermost beliefs.

As a philosopher overawed by science Rey did not see the
irony of his trying to prove from science the eternity of the
universe. The irony was all the heavier as no one in France had
until then studied Duhem's philosophical analysis of science in
a more systematic manner than Rey. More of this shortly. In
that analysis Duhem made much of the limitations of the
scientific method insofar as it was purely and strictly such and
not a mixture containing surreptitious borrowings from other
methods. Formulated by Duhem in a logically rigorous way,
those limitations applied to all particular scientific laws, even to
laws he did not specifically discuss, such as the laws governing
the kinetic theory of gases, an essentially statistical theory. It
was with an eye on that theory that Rey claimed the actual
periodic recurrence of the most unlikely configurations of any
group of material particles. The chief among those configurations was a cosmic state with a minimum of entropy.

According to Rey the statistical reinterpretation of entropy
disposed of its apparently chief implication: the eventual
inevitable heat death of the universe. Rey did not refer to
Duhem as he proceeded from the kinetic theory of gases to its
philosophical exploitation. He unabashedly spoke of a "metaphysics of eternal returns" as justified by physics. A revealing
but logical posture. Rey, fully aware of an essay written by
Duhem in 1893 on physics and metaphysics, had only one
choice: to be silent about that essay. There Duhem insisted,
among other things, on two points: metaphysics cannot give
specific guidelines for physical research which in turn he held
to be completely impotent to yield metaphysical conclusions.[5]

Rey must have had that essay very much on his mind as
something not to be brought at all into the picture drawn up
by him about an eternal universe standing for pantheism. He
had to pay a heavy price for this dubious strategy. It consisted
in a slighting of Duhem, the physicist, together with a rank
misrepresentation of what his physics was about. As he gave a
lengthy review of Carnot's principle, or the irreversibility of

thermodynamic cycles, Rey did not mention Duhem. Yet was it not Duhem who gave the first logically unobjectionable definition of irreversible processes? Did not Duhem's formulation of the thermodynamic potential represent a most important insight into the exact working of the law of entropy?

So much for the slighting. As to the rank misrepresentation of Duhem's thought, it came as Rey tried to weaken the significance of the Second Law of Thermodynamics by reviewing the principal philosophical interpretations attached to the notion of physical law. They were, in turn, pragmatism, subjectivism, and energetics. Duhem's thought had, of course, no connection with the first two.

In discussing energetics Rey analyzed the work of Selme, an unimportant physicist in the 1920s and for good reasons wholly forgotten today. According to Rey, Duhem's ideas were championed by Selme insofar as he wanted to construct "a thermodynamics along the pattern of geometry and rational mechanics, without any apparent appeal to experience." A page or two later Rey stated that according to Selme "the different quantitites of energy are the variations of the extensities through intensities and that it is their sum of products which according to the principle of the conservation of energy must remain constant in an isolated system." To this Rey added: "Here we recognize Duhem's fundamental method."[6]

To defend his speaking of Duhem in such a fleeting manner, Rey could have, of course, recalled that Duhem refrained as much as possible from making statements on his very subject, the entropy of the universe. Rey must have known that Duhem did not approve of such an application. Why then did not Rey invoke Duhem as a possible ally? Had he done so, he would have been forced to say something specific about the reasons for the position Duhem had taken. Those reasons did not relate to the debilitating paradoxes of an infinite Newtonian universe which were not discussed by Duhem or, for that matter, by Rey. Yet Rey should have

discussed those paradoxes, gravitational and optical, and should
have also discussed the new situation posed in cosmology by
Einstein's General Relativity, by then ten years old. Rey was
not the only "scientific" thinker in France at that time who felt
very much upset by Einstein's most emphatic claim that the
universe had to be finite. Was not the eternity of the universe
invariably tied, at least in modern times, to its alleged infinity?[7]

At any rate, Duhem's reasons related to rigorous logic
whereby he undercut attempts to attribute to physical laws,
however well established, more than their due. Rey who
wanted to establish scientifically the eternity of the universe and
give thereby the glamor of science to pantheism had no choice
but to say as little as possible of Duhem's views on the laws of
physics.

To be silent about Duhem was quite possible in France in
the late 1920s. By then the ideological left was very much on
its way to that takeover that came with the installation, in the
1930s, of the Front Populaire in the corridors of power, political and other. Rey did not risk provoking scholarly resentment
or rather the resentment of scholars in power by not saying a
word about Duhem, the historian of science. Rey's procedure
was all the more revealing as he functioned from 1920 on, not
so much as a philosopher as a historian of science, and in the
capacity of Director of the Institut pour l'Histoire des Sciences
et de Technologie. As a historian, and especially as a historian
of the idea of eternal returns, Rey had no scholarly justification
for ignoring Duhem's most scholarly treatment of the history of
that idea.

Once more Rey, for whom that idea meant the scientific
justification of his religion, pantheism, had no choice but to
give Duhem the silent treatment, the only escape for small
minds from the specter of a giant of the intellect. For a salient
contention in the work of Duhem, the historian of science, had
that idea for its target. Duhem claimed nothing less than that
modern, that is, classical Newtonian physics could not have

arisen if the Catholic Church, the only religious institution capable of doing so, had not disavowed the pagan cosmology of eternal recurrences. To ignore this claim of Duhem, which he supported with monumental documentation, and to ignore it in 1927 and in France, was part of that secularist ideological tactic for which, for many years now, the only way of coping with Duhem is to treat him as a non-entity.

In 1905 this silent treatment of Duhem was rather difficult to practice, although some had desperately resorted to it. Duhem's election in 1900 as corresponding member of the Académie des Sciences received world-wide publicity, partly because Gibbs and Boltzmann were honored in the same way at the same time. By 1900, Duhem, not yet forty, was a world authority in chemical thermodynamics, in hydrodynamics, and in viscosity. Furthermore, he proved his independence of thought as well as his keen critical sense by probing into the logically unsatisfactory aspects of Maxwell's electromagnetic theories.

It was difficult to ignore in France that Duhem enjoyed a growing reputation in the United States, especially at Cornell University, then the center of advanced studies in physical chemistry. It must have been a bitter pill to swallow for some in the Sorbonne, that even a doctoral student came from America to do research under Duhem's guidance in Bordeaux. It shows the measure of awareness about Duhem's pre-eminence in some unbiased scientific circles in Paris that in 1902 he was asked to write a series of articles on the evolution of the basic ideas of mechanics in the most widely read French scientific bimonthly, the *Revue générale des sciences pures et appliquées*.

The appearance of those articles, published shortly afterward as a book, *L'évolution de la mécanique*, established Duhem as a major philosopher of science. In 1905 Rey could not know yet that Duhem was soon to publish his great classic in the philosophy of science, *La théorie physique, son objet et sa structure*.

Within a few years Duhem's authority as a philosopher grew so strong as to force his antagonists to ask derisively whether the *Théorie physique* had become a sort of Bible, full of unanswerable arguments.[8] It contained in fact such cogently argued exposition of the limitations of the scientific method as to gravely undermine its use for ideological propaganda.

An even more telling blow to that propaganda was emerging in the measure in which Duhem proved himself a revolutionary historian of science. The preface which he penned on March 21, 1905 to the first volume of his *Les origines de la statique* contained nothing less revolutionary than a flat disavowal of a chief contention of those who prepared intellectually the French Revolution. According to them, it should be enough to recall Condorcet's *Esquisse d'un tableau historique du progrès de l'esprit humain,* reason, insofar it manifested itself in science, was man's only hope and that science could not arise until trust in the Christian scheme of salvation had been widely discredited. Renan proved himself but the latest link in a long tradition as he regaled his generation with his *Avenir de la science* that contained many variations of his worshipful encomium of science: "Science and science alone can provide humanity with that sense or symbol and a law without which it cannot live."[9]

To such blind trust in science as the savior of mankind, the chief conclusion of Duhem's researches on the origin of some basic laws of statics had to appear plainly sacrilegous. The conclusion called for nothing less than a complete recasting of the history of mechanics and by implication of the history of the laws of Newton, the very basis of modern or classical physics. For an age which took no cultural cliché so much for granted as the cliché about the darkness of the Middle Ages, the Age of Faith, nothing could be so upsetting as Duhem's claim that "the mechanical and physical sciences of which modern times were rightly proud, derive through an uninterrupted chain of hardly perceptible improvements from doctrines professed in the Schools of the Middle Ages."[10]

By March 21, 1905 Duhem was in the grip of what became the major concern of his last dozen years: the discovery and exploration of a vast intellectual continent, the medieval roots of Newtonian physics. As he hiked along the gorges of Tarn and of the Jonte in September 1905, his mind was less preoccupied with physics and its philosophy than with its medieval origins. His was an irresistible hunch that startling discoveries lay ahead of him. They indeed came much quicker than he expected. By early 1906 he knew of Buridan's formulation, and in a strictly theological context at that, of inertial motion. Two years later he came upon the discussion of Oresme of the rotation of the earth. Such facts spoke more convincingly than lengthy arguments about an intrinsic harmony between faith and science.

About these new pages of scientific history the champions of a "rationalist" historiography of science could do two things. One was a rather belated reaction initiated by Koyré in the late 1930s. It consisted in forcing a sharp wedge between medieval and Galilean concepts. Another, and far simpler thing to do was to avoid facing up to the evidence about medieval science. This is what Rey did. He was long dead when the reading of Duhem's historical studies on the evolution of the basic concepts of mechanics prompted C. Truesdell to speak of Duhem as a genius compared with any other highly regarded present-day historian of science.[11]

A genius he was and a most disturbing one for a variety of intellects. The ones that felt most uneasy about him were kindred souls to Rey. They all staked their secularist ideologies on a mechanistic interpretation of nature in terms of Newtonian physics which represented for them the highest and only reliable form of rationality. Implied in this view was the discrediting of metaphysics and all that was meant by its cultivation. An ideological overtone rang out in the declaration which Cornu, a prominent French physicist, made in 1900 at the International Congress of Physics in Paris: "The more we penetrate into the knowledge of natural phenomena, the more developed and precise is the audacious Cartesian conception of the mechanism of the universe."[12]

The champions of mechanistic ideology had at best a condescending smile toward those who kept asserting the intellectual respectability of acknowledging supernatural reality, especially in the form of miracles. "By its principle as well as by its conclusions, science excludes miracles," declared confidently in 1903 Gabriel Séailles, a chief ideologue in the Third Republic. Those were the years that reverberated with Renan's dictum: "Science will organize God himself." Soon that overweening confidence in science was to receive its proper label, scientism, in the hands of Maritain, once himself a devotee of scientism.[13]

More threatening than appropriate labels, whose effectiveness must not be understimated, had to appear the fact that the conceptual foundation of mechanistic science had been by 1900 or so the target of an increasingly large number of scholars who certainly could not be suspected of lack of expertise about physics or lack of philosophical acumen. Among these Duhem was a towering figure. Rey served evidence of this when in 1907 he published his major philosophical work, *La théorie physique chez les physiciens contemporains*, a book of more than 400 pages.

As a convinced ideological mechanist, Rey felt that mechanistic philosophy still could serve as the foundation of all rationality. This he set forth in the second part of his book. In the first part he surveyed, in turn, the hostile critics as well as the plain critics of mechanism, followed by a summary of the works of physicists still believers in mechanism, whom Rey saw as the wave of the future. Poincaré was the chief among the plain critics. His idea that the laws of physics were mere commodious devices to organize data posed, so Rey thought, no serious threat to mechanistic realism. Material reality had its own built-in credibility for anyone appreciative of common sense.

Among the hostile critics were Rankine, Ostwald, Mach, and Duhem. Rey gave far more attention to Duhem than to any of the three others, although Ostwald and Mach were much more widely spoken of than Duhem. Perhaps Rey realized that Duhem articulated his case against mechanism with far greater cogency and lucidity than those two. More revealingly, Rey just did not seem to know into which philosophical category or school to put Duhem. For the articles and books Duhem had until then (or even afterwards) written on physical theory could not be fitted into the philosophical moulds discussed at that time by professors at the Sorbonne or other secular places of learning.

Nothing was, of course, easier than to separate Duhem from those physicists, and there were still many of them around in the opening years of the 20th century, for whom physics was the tool that revealed the machinery of the universe. Shortly before Duhem began to write on the philosophy of physics, he had recognized some self-defeating philosophical presuppositions of mechanistic or Newtonian physics. But a non-mechanistic or positivist idea of physics was endorsed by Duhem with so many qualifications as to make it impossible for Rey to classify him as a positivist. Much less did Duhem represent that extreme form of positivism which Poincaré held and advocated under the label of commodism.

While Duhem provided some clues about the true category to which his philosophy of physics belonged, he did not elaborate them and much less did he put on his philosophy the only label, Neothomist, appropriate to it in ultimate analysis. Nothing, in a sense, would have been farther from Duhem than to be associated with Thomas Aquinas, about whom he knew at that time very little, and whom he was to charge, a decade or so later, with the greatest failure, inconsistent thinking, that can be laid at the door of a philosopher, let alone a great philosopher. Yet already in 1893 Duhem endorsed metaphysics, together with a realist epistemology, as the foundation and completion of all knowledge, scientific or other.[14] This was an unmistakably Thomist position, recognized as such by some Neothomists whose leader in French-speaking countries was Mercier, the future cardinal archbishop of Malines.

Such a perception of Duhem would have been impossible in those secularist realms which resounded with Harnack's battlecry: "Catholica non leguntur." The Sorbonne, or at least some at the Sorbonne, learned about Neothomism when Père Rousselot defended there in 1912 his famous thesis, *L'intellectualisme de Saint Thomas d'Aquin.* One wonders whether Rey, or thinkers moulded in the identification of the rational with the mechanical, would have looked up the book of Père Rousselot had he come up with it a few years earlier. Even longer treatises would not have alerted them to what Duhem had already claimed in brief but incisive paragraphs: Plain recognition of simple facts and the registering of obvious objects do not cease to be rational just because they are mental acts presupposed by physics instead of being justified by its method, mechanistic or not.

Such was the gist of Duhem's affirmation of the indispensability of metaphysics. Insofar as they were intelligible, plain objects and plain facts were provided by a metaphysical act of the intellect for the physicist bent on investigating their quantitative aspects. This, for Rey, meant however the steeping of physics in foundations that were not an object of reason but of some quasi-mystical attitude, a sort of faith. Hence Rey's characterization of Duhem's theory of physics as the physics of a believer. He did not mean thereby Duhem's ardent Catholic faith which or may or may not have been known to him. Although by 1904 Duhem had published a dozen books and a hundred or so articles, he never referred in them to the Catholic faith, let alone to his own profession of it.

For all his acquaintance with many Thomists, Duhem never grew fond of Thomistic terms. Had he done so, he might have perceived the misconception of the term "faith" as a label put by Rey on his ideas about what a good physical theory ought to be. Instead, Duhem took Rey's use of "faith" in a theological sense and replied in kind. As one walks in an argumentative mood, one can easily miss the immediate countryside and even walk well beyond one's chosen target. Duhem certainly missed the point Rey wanted to make. It is to that misunderstanding on Duhem's part that we owe his magnificent profession of Catholic faith in a scientific context. He lived that faith from childhood to sudden death in his mid-fifties with heroic fidelity and with a profoundly reasoned conviction.

Yet that misunderstanding of his contains in a nutshell a most instructive aspect of his being a believer as well as a scientist. The misunderstanding was that of a philosopher of physics, but it was his overriding interest in a perfect form of physics, a heavily philosophical objective, that prompted him to unfold with great originality previously unsuspected harmonies between faith and science. The science was physics, which even more than in his time stands, because of its exactness, for the ideal to be emulated by all other sciences. Though a theoretical physicist, he deeply appreciated the role which industrial applications played in the progress of his field.[15] The depth of his views about the abuses of physics for destructive purposes can be gauged from his remark that the ravages of World War I constituted that gravest of all sins which is the sin against the Holy Spirit.

Here too he was led by a faith that had for its object the teachings of the Catholic Church which Duhem followed and practiced with no mental reservations. Even in most difficult situations, which State as well as Church created on more than one occasion around the turn of the century, he did not let his personal preferences override his sense of loyalty to Church authorities. Nor did he grow resentful when in vain he tried to call the attention of those in charge at the Institut Catholique in Paris to the decisive role which studies in the philosophy and history of physics were to play in the shaping of twentieth-century cultural consciousness.[16]

Catholic institutions of learning still have to do justice to that role. They might put themselves into a commanding position if they recognized Duhem as one of the greatest and most reliable geniuses of this century of science. Those teaching in Catholic universities can always derive inspiration from his fearless courage whereby he kept disregarding the interests of his own academic career. The one who trusted so much in Divine Providence offered no hollow rhetoric when he portrayed, in his *Les Origines de la statique,* unquestionably the most revolutionary work in the historiography of science,[17] even the evolution of science as something directed from above. Apart from his devotion to science, his life too was, from start to end, a witness to a faith whose hallmark is trust in the ultimate victory of virtue and truth.

2

Prelude

That Pierre Duhem lived his faith through his entire career is in no small measure due to the example set by his parents. His father, Joseph Duhem, came from Flanders to Paris in 1852 at the age of twenty-three and in 1859 married Mlle Marie Alexandrine Fabre, descendant of a family with long ties in Cabrespine (Aude). Pierre was their first child, born on June 9, 1861, followed two years later by twin-sisters, Marie and Antoinette. The family lived at 42 Rue des Jeuneurs, a street then as now full of textile shops, in one of the modest apartments that could be afforded by Joseph Duhem, a textile salesman.

Joseph Duhem was much more than a breadwinner. "Under a quiet and reserved exterior, he hid a great soul, all goodness and tactfulness," so was he recalled by his granddaughter.[1] For Joseph Récamier, renowned physician and for a while an associate of the Duc d'Orléans on his Arctic expeditions, Joseph Duhem "was a plain and good man."[2] Récamier, himself an exemplary Christian, who as Pierre's friend from boyhood often met Joseph Duhem, must have, as he penned that short phrase in 1933, meant to convey more than purely human characteristics. In the Duhem home it was easy to resonate to Christ's words about "the faithful and prudent servant," applied in the liturgy to Saint Joseph, that most unassuming model of heads of all Christian families.

No different was the case in connection with that Mother who was the heavenly patroness of Pierre Duhem's mother. His devotion to his mother was proverbial. Colleagues of Duhem at the University of Bordeaux and friends of his among the clergy were time and again startled by the great *savant's* deference to his mother's wishes and views. The fourth commandment, Duhem used to reply, "doesn't say that a mother, once old, is no longer a mother! . . . Moreover, one has one's mother only once. Were I to stop obeying, it would seem to me that I had lost my mother."[3]

One can easily picture Mme Duhem taking her son to weekly catechism at St Roche, a half-an-hour walk from the Rue des Jeuneurs. Young Pierre learned from that weekly obligation that religion is above all a commitment to a service imposed from above. At St Roche he saw that commitment exemplified in the courageous comportment of parish priests during the Commune. He was one of the six intrepid boys who kept up with the duty of going to catechism even in most difficult days. His reward, a medal from the pastor, the abbé Millaud, came when, with the Commune over, the abbé told the story — partly hilarious, partly suspenseful — of his escape from revolutionaries who wanted to arrest him in his apartment.

Religious sense of duty was a hallmark of the Duhem home. Young Pierre could not fail to be impressed by the reason why his father did not go to lycée, let alone to a university. The oldest of seven children, Joseph Duhem had to leave secondary school at the age of fourteen, when he lost his father, to help his mother provide for the large family. Young Pierre was reminded of one's duty to sacrifice one's own interests whenever he saw his father go out for a walk holding in his hand the book of a Latin author. Education had to be seen by young Pierre as a gift from Providence and not a right that one could claim under any circumstance.

As the Prussian troops approached Paris, Joseph Duhem, like many other Parisians who had relatives in the South, took his wife and children out of Paris. They went to Chateaudun where a cousin of Mme Duhem served as the state attorney. As the siege of Chateaudun reached its final phase in mid-October 1870, Joseph Duhem found shelter for the family in the basement of the hospital as a protection against artillery fire. Then he went out to see whether it was still possible to move outside the town. He was on his dangerous errand when the refugees in the basement had to choose between possible suffocation or an almost certain exposure to gunfire. Pierre's mother was one of the few courageous refugees to exit and seek shelter elsewhere as bullets whistled across the streets.

Young Pierre's letter about the siege, written a few weeks later to an uncle in Paris, is a marvel of precision and factuality. Printed here for the first time in English,[4] it should reveal much of the frame of mind of the great *savant* the young boy was to become. The courage of that great *savant* was foreshadowed in the fact that he was not at all terrified as a bullet came through the hospital basement window and lodged in the wall next to his head. He removed the bullet from the wall and kept it for years as a souvenir.

Once the family was reunited just outside the town, they had to be ready in the middle of night for a dangerous and exhausting march, described by Pierre in vivid detail in the same letter. Finally they reached Bordeaux. From there they returned to Paris shortly before the Commune began in March 1871. In the final week of the Commune Pierre could see from the Rue des Jeuneurs the flames and smoke coloring the sky as the Tuileries and the Palais de Justice were burning. What affected him most was the desecration of Notre-Dame-des-Victoires, a famed church not too far from his home in the second *arrondissement*. As an adult he must have recalled more than once what he saw: the statue of the Virgin dressed up in the garb of a revolutionary, a detail vividly recalled by Hélène, his only child and first biographer. She was also the one who kept for posterity the information that as an adult, and living far from Paris, Pierre Duhem did not fail to make a brief visit to that church whenever he traveled to the capital.

The desecration of his own parish church, Notre-Dame-de-Bon-Secours, took place in a style that recalled the worst that happened to many churches during the French Revolution. As a child of parents in modest circumstances, Pierre must have had his share of the hunger that during the last days of the Commune was a greater threat to life than wild sprees of arrests and streetfighting. He must have heard revolting details about the unconscionable behavior of some food merchants and of the few with enough money to satisfy their greed. His future nemesis, Marcelin Berthelot, was among those *bons vivants*.[5] The seeds of a lifelong compassion for the hungry were planted right there and then in Pierre's soul which was protected from turning into a resentful reformer both by his faith and by what he saw. Forty three years later, in a letter to his daughter, by then a social worker in Paris and flirting with reformist ideas, he began his reminiscences with lapidary remarks: "March 18. Anniversary of the Paris Commune. You have not seen it. Paris is in the hands of bandits." Subsequent lines from that letter belong to the chapter on Duhem the Christian.

The Commune had been a memory for a year or two when Pierre heard about a quiet Catholic hero of the harrowing time that began when the Prussian cannonballs started hitting the City. He was the abbé de Lagarde, who a few years earlier, at the age of 37, took over the directorship of the Collège Stanislas. While he kept the classes, he also made room in the Collège for the injured, in addition to giving spiritual solace to them. Long before these details appeared in print, word about them must have made the rounds among Catholics in Paris. It was not an accident that Pierre's mother chose, following a visit with the abbé de Lagarde, Collège Stanislas as the place for her son's education. As one who practically laid new foundations for Stanislas, the abbé de Lagarde chose for the school's motto: "Frenchmen without fear, Catholics without reproach."

Mme Duhem was not disappointed in her expectations. The abbé de Lagarde had been dead for twenty years when in 1904 Duhem was asked to contribute to a book in which distinguished alumni of Stanislas recalled their memories of the school. Duhem, already a *membre correspondant* of the Académie des Sciences, wrote about the Ecole Préparatoire of Stanislas, famous for its success in placing its graduates in the *grandes écoles:* Ecole Normale Supérieure, Ecole Polytechnique, and Saint Cyr. His account began with with some hilarious aspects. Just as now, in the 1870s too, students were ready to take full advantage of the weak points and foibles of teachers, admired as they were otherwise. Duhem's part in those tricks consisted in making of this or that teacher speedy caricatures, some of them confiscated on the spot and pocketed by their victims with ill-concealed satisfaction. His ability to draw was legendary in the school. When in 1884 the abbé Lagarde died, the new director of the school called on Pierre, already at the Ecole Normale, to come and aid with his sketches of the abbé Lagarde the famed sculptor, Henri Chapu. The latter was commissioned to make of the deceased director a marble bust which is now the chief décor of the Salon Rouge of the Collège.

After he had dealt with the hilarious details, Duhem recalled the seriousness of studies. Stanislas did not cater to half-measures, let alone to laziness and ineptitude. Had Duhem not written more about his four years in the Ecole Préparatoire, his account would already have turned into a classic never to be forgotten by future students looking there for camaraderie and intellectual formation. But there was something even more important at Stanislas. Owing to his deeply lived faith, young

Pierre Duhem had eyes and appreciation for the evidence of holiness. His gripping portrayal of the cancer-stricken abbé de Lagarde's heroic comportment comes to a close on a note of abiding gratitude:

> He knew each of us better than a father knew his children. He bestowed on us, with a surprising and supreme solicitude, all that strength which that ravaging sickness left for him. When the abbé de Lagarde spoke to us, we were troubled to the depth of our soul because we felt that we had just talked with a saint.[6]

In addition to the abbé de Lagarde, Pierre met at Stanislas other ecclesiastics, so many living rebuttals of the words, HERE LIES THE LATE FRENCH CLERGY, which students entering the school could not help noticing on the facade of a former Carmelite cloister just across the street. The inscription, put there by brave champions of the Revolution, was an ironic commemoration of the many priests and religious sentenced to death in the *brasserie*, now a part of the school's entrance, and executed in the cloister. A living refutation of the clergy's demise was none other than Mgr Ségur. Pierre's first encounter with that saintly prelate came while preparing for his first confession in October 1872. On returning home, he announced triumphantly: "I picked the bishop!"[7] The bishop, who was one of the several eccleasiastics available for the students, soon became the spiritual advisor of Pierre's mother. Pierre is listed in the *Annuaire* of the Collège for 1872-73 as one of fifty-nine boys who took their first communion from the hands of Mgr. Ségur on May 21, 1873.

Something from Pierre's lifelong devotion to the Eucharist can be gathered from the fact that he had carefully kept two small items. One is a card commemorating his first communion, the other is a printed list of prayers and hymns, recited at the celebration of the Feast of Corpus Christi at Stanislas. One can readily assume that he was a part of the yearly pilgrimage of the

Collège to the Sacré Coeur on Montmartre whose construction began in 1873. It was to serve as a monumental reminder for the need of a spiritual renewal for France in the wake of the humiliating defeat of 1870. The theme of *Gallia penitens*, embodied in that church, found vivid resonance in the soul of Pierre who, whatever his ardent patriotism, always kept an eye for that greatest threat to any nation which is spiritual and moral decay.

The contribution of Pierre's years at Stanislas to his eventual turning into a signal representative of the unity of faith and science has further important aspects. A devout boy he certainly was but not one who ever thought of becoming a priest. This is all the more significant because he had warm relations with fellow students who were to excel as missionaries in Indochina. When two decades later news from there about religious persecutions reached metropolitan France, Duhem must have often spoken of a former *camarade* of his in the presence of his daughter. Otherwise she would have hardly remembered the name of Jean de Guébriant, who, in all evidence, suffered a martyr's death.

No ecclesiastical vocation was sparked in Pierre's soul by his devotion to the abbé Pautonnier, a young instructor at Stanislas. Thirty years later he accompanied Duhem on a hiking trip in the Pyrénées. Possibly he was the one who saw Duhem caught there by the Spanish border guards, escape from their clutches and poke fun at them from the French side. Another and still young priest, the abbé Biehler, director of the Preparatory School, might have represented a far more tempting reason for Pierre to think about ecclesiastical calling. The reason is embodied in the words, PRO DEO ET SCIENTIA VIXIT, which crown the abbé Biehler's bronze bas-relief in the wall of a building in the School's courtyard. While his ecclesiastic collar is a plain reminder of his service to the Church, few knew at Stanislas, let alone elsewhere in France, that he was a prominent student of the famed mathematician, Hermite, who in his elder days kept visiting him at the Collège.[8]

When a new boy at the College, who smiled on seeing an old man, limping on a huge cane, make his way to the abbé Biehler's small upstairs room, an older boy whispered into his ears the name of that handicapped old man: "Hermite!" The older boy must have been Pierre, one of those very few whom the abbé Biehler called in to the presence of Hermite. The one, to quote Duhem, "whom the entire scientific Europe viewed as the purest incarnation of the mathematical spirit," wanted to meet budding mathematicians. The rest of Duhem's reminiscences of Hermite's visits at Stanislas prove by their vividness that he described himself in the student who upon returning "to the study hall after one such interrogation . . . had the awareness of having seen at close range a genius."[9]

It would have been natural that such connections with the clergy would put the seed of ecclesiastical calling in the mind of a young man intent on the highest moral and intellectual ideals. Pierre, most serious about his faith, showed great promise in mathematics and physics, and showed little interest in the fair sex. Most importantly, he had that marvelous quality, unaffected humility, that always makes the most efficient priests. Its irresistible impact made itself felt long beyond those school-years and even beyond Duhem's life. He had been dead for over a dozen years when his friend, Joseph (Joe) Récamier, recalled, from a distance of fifty years, the indelible impression which precisely that quality of Pierre made on him:

> He certainly was conscious of his intellectual superiority. It was impossible that it should be otherwise. Yet, throughout our youthful years he never said to me, who lived so closely to him and was so inferior to him, a single word that might have hurt me in this respect.[10]

Joe Récamier, a retired physician by 1932, had as a young man observed Pierre often and at close range also in the

Duhems' modest apartment. Whether Pierre was still at Stanislas or at the Ecole Normale,

> he was a most respectful son towards his parents, the most deferential son I have ever seen. His father represented, I believe, producers of textiles. He was a simple and good man whose career was modest. He lived in a small apartment in the Rue des Jeuneurs where I often visited and where I assisted him at his deathbed. Pierre Duhem, in the midst of his own family, sitting next to his young sister, was modesty itself and I wondered at times why his father did not take more pride in his son whom his teachers had already declared to be exceptional.[11]

He was exceptional in more than one respect. He could have become a first-rate classicist, a lawyer, an artist, and even an actor, and in general ready to espouse any good cause. But already in the Ecole Préparatoire he felt that he should devote his whole life to the cause of physics. His fellow students kept repeating with awe and admiration a remark of Moutier, professor of physics at Stanislas, that they should watch their *camarade* Duhem because he was destined for making discoveries.

Most importantly, Pierre took an exalted view about physics. The perfection or dignity of physics had in his eyes a sacred character, not to be tampered with by incompetent ecclesiastics, however well-meaning. More of this later. Physics was not even to be joined in a "holy alliance" with religion. Duhem's mature view of the relation between science and faith had for its pivotal point the recognition that the two had different objectives and methods. Precisely because of this, physics could be turned against faith only by turning it into a pseudo-metaphysics, that is, into a vehicle of a "lay-religion." The latter had, in the heyday of the Third Republic, more than one notable champion, all eager to exploit physics for the purposes of materialism, crude or sophisticated.

A view of the true nature or limitations of physics must have intrigued Pierre from his years in the Ecole Préparatoire of Stanislas. Otherwise he would not have achieved, early in his teaching career a clear idea about what physics was. Coupled with his enormous esteem for the beauty of physics, that idea became the guiding principle and chief motivation as he worked out priceless contributions to a relation of faith and science in which they were united with no detriment to their unity as well as mutual autonomy.

It was in that very deep sense that Pierre chose physics as his life-calling. It was a vocation first in the sense that it offered no ready prospect for becoming prosperous. That prospect might have been better served had Pierre chosen the Ecole Polytechnique. Such was the suggestion of some industrialist friends of his father. They hoped that the talented young man would eventually make himself available to them. No less importantly, Pierre's lay-vocation had no overtones of laicism, not even the overtone of a dichotomy, fashionable then as now, between faith and knowledge. Rather, it fully represented the age-old Christian conviction that all created things were intrinsically good and their study worth pursuing for their own sake. While there was no such a thing as a "Christian" physics, physics as truth had to be part of that truth which, if it was truly one, had to be universal, that is, Catholic.

Such considerations formed the basis of a lifework which, without ever being intended to be an apologetics, issued in first-rate and, in fact, epoch-making contributions to the defense of Christian faith and culture. The same work could therefore become a means of personal sanctification without a need to mix it at every turn with devout thoughts. Considerations of this kind were merely implied in the education Pierre received at Stanislas. There the Catholic view of truth was rather practiced than taught or preached. The fideism, so dear to the abbé Gratry, director of Stanislas in the 1840s, and to many in France then as now, failed to set a pattern there as did other trends, including Thomism. When the latter began to be voiced by Maritain, a teacher at Stanislas in 1911-14, some parents wondered whether exposure to Thomism might not diminish their children's chances to be accepted in the *grandes écoles*.

A somewhat pragmatic freedom of thought had prevailed at Stanislas even when Pierre studied there. The principal history teacher was Louis Cons, famed for his textbooks as well as for his advocacy of that part of Comte's positivism in which the study of the history of a subject formed a pre-condition of its true comprehension.

This view of the study of history was a seed that found a most receptive soil in the mind of Pierre who at one point felt tempted to become a historian rather than a physicist. No such temptation was posed to him by the study of Latin and Greek classics. His mastery of both, together with his fondness for history, proved to be an indispensable tool for his eventual discovery of an unsuspected intellectual continent, the pre-Galilean, that is, medieval roots of Newtonian physics.

In spite of his intellectual excellence, Pierre did not aim at obtaining the best marks. He wanted to have time for special studies, which he pursued on his own. He had a passionate love for biology. The free Wednesday afternoons often found him in the Museum for Natural History. He loved to visit the Louvre and the Musée du Luxembourg which at that time contained the permanent collection of modern, that is, 19th-century French paintings and sculptures. Time and again he went on foot to Versailles on Sundays, accompanied by his best friend, Joe Récamier.

During those long walks the conversation must have often turned to topics of religion. The trend in the newly born Third Republic took on ever stronger anti-clerical overtones. Free-thinkers had an ascendency that could not pass unnoticed by Pierre and Joe who in their homes developed a keen sense of loyalty for the Church and historic French Catholicism. Fifty years later, in a letter to Hélène, Joe recalled the admiration he had felt about Pierre's well-reasoned defense of Catholic viewpoints and positions.

Obviously, Pierre learned a great deal about religion and Church at home. During his years at Stanislas it was at home that he received a most important training in religion which could not have started on a happier note. On September 20, 1872, Pierre served as the godparent of his newborn brother Jean. Two months later, in a particularly cold and humid November, the little baby fell victim to laryngitis that raged as an epidemic all over Paris. Pierre and Marie were rushed to relatives. Antoinette, the other of the twin sisters and already sick, fought for life at home, but not even a surgical intervention could save her. As one who received first communion before her death at so innocent an age, she was remembered in the Duhem family as their angel and all the more so as her death occurred on November 24, the feast of Saint Catherine, virgin and martyr.

To accept those two deaths with trust in God was for Pierre a hard training in faith. It brought along other lessons in religion. One was a reminder of life's fragility. In that respect young Pierre received further promptings from the serious rheumatism of the stomach he had contracted in 1874 during a vacation in St Gildas de Rhuys, on the Gulf of Morbihan. He was sent there after a doctor recommended exposure to sea air. The doctor did not take into account the avid biologist that was young Pierre. His search for marine life in the tidewater had for one its results that intestinal rheumatism which kept him out of school for much of the year 1875-76 and, to his life-long sorrow, made him unfit for military service. A minor victim of those vacations at St Gildas de Rhuys was Pierre's luggage which he carried home stuffed with molluscs, crabs, starfish, and shells of all sorts.

This last detail was preserved for posterity by Marie, a chief beneficiary of the extra attention Pierre showed after those two deaths that struck the Duhem home in quick succession. For Marie he tried to be a replacement of her twin sister. From that time on he showed an even deeper consideration for his parents, especially his mother, who for the rest of her life failed to recover fully from that double tragedy. Yet it was from his mother that he received a most needed spiritual support when toward the end of his stay at Stanislas he got his first bitter taste of a most unsavory aspect of academic life. One of his teachers there published under his own name something that Pierre had noted and worked out in the first place. To cope with the frustration Pierre's mother offered her share in practicing patience and humility, both of which stood him later in good stead on seeing brazen plagiarisms of his researches. Jean Perrin, the future Nobel-laureate, was to earn his scientific wings by publishing books on chemical thermodynamics without giving as much as a hint of the enormous debt he owed to Duhem's publications.[12]

Other tragedies were not to spare Duhem whose life seemed to be destined to a long series of triumphs from the moment he entered in the Fall of 1882 the Ecole Normale at the head of his class of about forty, chosen from over 800 new graduates from lycées all over France. In many ways his life in the Ecole reflected the happiness of an extremely talented young man confident of the splendid future ahead of him. His intellectual superiority did not prevent him from being most accessible and helpful to younger *camarades*. None was more grateful for this than Jacques Salomon Hadamard, two years his junior and a friend for the rest of their lives. Their friendship would have hardly lasted if Duhem, an ardent Catholic and a Royalist, had resented anyone just because he was a Jew, or a liberal, or even a Dreyfusard.[13]

Duhem was the life of the Ecole through his ability to mimic others, his readiness for practical jokes, and his flow of caricatures. He knew how to be an integral part of an in-creasingly secularist atmosphere, without compromising his religious convictions, let alone being ostentatious about them. His moral seriousness had, as Joe Récamier pointedly recalled, no touch of prudery. He was one of the *talas,* that is, one of those Normaliens who regularly went to mass on Sundays to the nearby Saint-Jacques-du-Haut-Pas. But he did not join such Catholic works as the Conférence Saint Vincent, although many *talas* were active in it. Neither then nor later was he in sympathy with activism.

Yet, when he felt that it fell to him to respond with action to an eminently Catholic duty, he rose to the occasion. One such case is connected with the early death of a *camarade* of his at the Ecole Normale, Bronislav-Etienne Wasserzug, a most promising assistant of Pasteur. Following Wasserzug's death on March 30, 1888, the editor of the *Annales de l'Institut Pasteur* recalled that Wasserzug was the first to study in laboratory the rate of transformation of species, centering on various mushrooms. Because of his friendship with Bronislav, Pierre, already in his first teaching job at the University of Lille, was asked by the Association of the alumni of the Ecole Normale to write the obituary.

That obituary would deserve to be recalled if only for the beginner teacher's courage to expose the insensitivity of those in power in Paris towards Bronislav's father. Many among the same could be but incensed by Pierre's courage to give the center place in that obituary to Wasserzug's slow recovery of his Catholic faith and his exemplary death as a Catholic. It was with Christian modesty that he concealed as much as he could his own role in his friend's conversion. Only those ready to read carefully between the lines could guess the priestly role of laity which Pierre discharged to the benefit of his friend, without trying to slight the role of priests. He knew how to intimate much by the unaffected reference to the fact that it was on Good Friday noon that his friend's "heart, which resonated with so many noble affections, beat for the last time."[14]

By the time these lines saw print, Duhem had already begun another form of lay apostolate: his life-long support of Bronislav's widow, only twenty and the mother of a three-month-old son when she lost her husband. In general, Pierre's reluctance to get involved in organized charities, let alone in other types of student activism, reflects his concern about the proper use of time. While at the Ecole, his sole recreation consisted in quick Sunday-trips to Argenteuil on the Seine where his friend Joe Récamier, a student of medicine, kept his small sailboat. During the summer break he went in Joe's company to the Breton seacoast where the two improved their sailing skills. The scenery also inspired a series of Pierre's sea-scapes, some in aquarelle, most of them in India ink. Apart from this Pierre's was a total dedication to his life-goal, the working out of a perfect form of physics.

Far from being unnoticed, that total dedication to a most specific goal made a deep impression on precisely those among his fellow students who later themselves excelled as scientists. One of them was none other than Hadamard who from a distance of forty years offered the following recollection about Pierre, the accomplished mathematical physicst he had been already as a Normalien:

> He was a physicist and a physicist he wanted to remain. It is well known that this vocation of his did not have to wait for the Ecole Normale to assert itself. For us his camarades in the Ecole, this precocity of his was not the only subject of astonishment. The taste for physics was rare at that time when, it must be spelled out, we felt around us something of a stagnation with respect to that science. How marvelous, in contrast, was our enthusiasm for mathematics when facing a Hermite, a Poincaré, a Darboux — to speak only of those already dead — and when we were in the serene and inspiring company of [Jules] Tannery. No one felt this enthusiasm more completely and profoundly than Duhem whose knowledge had been truly

universal and who, as is well known, could just as well
have become a biologist as he could a mathematician and
a physicist.[15]

A similar testimony came from Pierre's classmate, Pierre
Houllevigue, who in 1893 became professor of physics at the
University of Montpellier:

> Duhem was my *camarade* at the Ecole Normale. He was the
> *cacique* [student leader] . . . and what a *cacique*! When we
> entered the Ecole as students still almost green-horns,
> rough sketches of the men we were to become, Duhem
> was already a fully developed man. His character and
> intellect had already taken their definitive form. He knew
> what new truths he would bring to the world. In fact, he
> was already a master and we, who alongside with him, had
> not for a moment the idea or the wish to contest his
> intellectual superiority.[16]

For all his single-minded devotion to his chief purpose, he
was no narrow-minded specialist. Emile Picard, one of his
teachers in mathematics, who concluded his brilliant career as
Perpetual Secretary of the Académie des Sciences, recalled
Pierre's years at the Ecole as "the happiest in his life," and
added:

> No one was less a man of a single interest than Duhem,
> and already then his amount of reading was immense. This
> manner of studying is not always the most favorable for
> succeeding in the exams; these have their hazards, and
> Duhem was ranked seventh in his exams for the licenciate
> in physics and chemistry, but he was left at the head of his
> class, so evident was his superiority.[17]

No wonder that from the very start his stay at the Ecole
pointed towards a splendid future. Already at the beginning of
his second year there he was heading toward his first major

scientific discovery, the formulation of the idea of thermodynamic potential. During the winter 1883-84, several of his
papers were read at the Académie des Sciences under the
sponsorship of Darboux, a leading chemist and member of the
Académie. Soon afterwards he was encouraged to work out his
finding in the form of a doctoral dissertation. He so much impressed his teachers, that — although not yet an *agrégé*, not
even in the possession of licences, and not even a graduate of
the Ecole Normale — they obtained for him the permission to
present his doctoral thesis at the Sorbonne.

Then in June 1885 tragedy struck. His thesis, an indirect
rebuttal of Berthelot's maximum work principle, brought down
on him the bitter resentment of an academic establishment in
which too many owed too much to Marcelin Berthelot. An
excellent experimental chemist, Berthelot was a deplorable
theorist. Worse, he felt he was called to act as the chief guru
and power broker of the French academic world. At one point
he even served as Minister of Foreign Affairs. Last but not least
he was a zealous Freemason of the militant Gallic brand.

In Berthelot, already a member of the Académie des
Sciences and soon to become its Perpetual Secretary, Pierre
found a chief example of those intellectuals for whom the
search for truth is largely a matter of lip-service to their
professed respect for it. Their real interest lies in the pursuit of
their own vainglorious reputation which they are ready to
protect at any price. In Pierre's case the price consisted in
Berthelot's persuading Lippmann, a future Nobel-laureate for
his work on color photography, to reject Pierre's thesis on
patently flimsy grounds.

One of the greatest French physicists of the turn of the
century, Pierre Duhem was denied a doctorate in physics. He
was in fact so great already at that time that within six months
he completed another thesis, in mathematics, which he defended with the highest honors before a jury that included no less
a mathematician than Poincaré. His rejected dissertation, now

in the Microfilm Landmarks of Science, was printed at Duhem's own expense in 1887. By the time it came out in second edition in 1897, Pierre Duhem had three further tragedies to carry on his shoulders. One of those tragedies centered on his professional career, a purely lay-career on a first appearance but with deep relevance to the relation of science and faith.

3

Career

It was the normal course of life that Pierre Duhem should lose his father whose death came on April 7, 1899. One of his consolations was that his mother could now move, together with his sister, to Lille where he was at that time completing his second year of teaching as assistant professor. He arrived in Lille in late October 1887, just after the transfer of the Faculties of Letters and Law of the old University of Douai to Lille had been made official. It was a move bitterly disputed by the citizens of Douai who had already strongly protested in the 1850s, when the central government in Paris set up in Lille a Science Faculty with a legal status as a branch of Académie de Douai (French provincial universities remained mere "Académies" until the early 1890s).

Unlike in Douai, where all citizens opposed the gradual transfer of their university to Lille, the citizens of Lille were divided. Those with anticlerical sentiments had vested interests in benefits coming from the government in Paris that from the 1880s on pursued militantly anticlerical programs. They, of course, favored the establishment of a full-fledged university which, in the words of Jules Ferry, was to be "a citadel against a citadel in the vast field of liberty." The other citadel was the "Catho," or the Institut Catholique, the best established of such

Institutes in all France, then as now. It grew up within ten years into such a powerful place of learning, with a Faculty of Medicine so outstanding, as to pose not a few headaches to the ideological champions of the new State University. The latter began a vast program of expansion after Berthelot became President of the Conseil Supérieur of Public Instruction in January 1887. It was not without prompting from higher places in Paris that the editor of *Progrès du Nord*, the leading Republican daily in Lille, found it opportune to editorialize shortly afterward: "If the Faculties of the State University had formed a tight unit ten or so years ago, instead of going their own ways, the clericals would have hesitated before creating a Faculté Catholique. The reuniting of the Faculties of the State University is a must if one is to fight successfully the teaching given under the auspices of clericalism."[1]

That the government poured into the newly founded University of Lille money and personnel to assure its high standards could be of some comfort to Duhem for his not having been given a post in Paris in one of the *grandes écoles*. Paris was also much more accessible from Lille than from Bordeaux, let alone from Toulouse, seats of the two most eminent provincial universities in France at that time. However, he could not help realizing that he was entitled to a post in Paris in view of the volume and quality of what he had published by the summer of 1887, when he also finished an exceptional second year in pure research at the Ecole Normale. Yet when he was assigned to Lille as his first teaching post, he could hardly have realized in full the ominousness of Berthelot's informal decree to the Ministry of Public Instruction: "This young man shall never teach in Paris."[2]

What became an obvious loss for students of physics in Paris, turned into an unsuspected boon for those in Lille. The yearly reports on Duhem that were sent by his superiors over almost thirty years to the Ministry of Public Instruction form a thick dossier, now in the Archives Nationales. It contains many follow-ups to the reports of Couat, rector of the University of Lille, who in 1889 and 1890 informed the Ministry:

> Since my arrival in Lille [in 1887] Duhem devoted himself to his duties with extreme zeal. Neither the vast amount of his personal research, nor the poor condition of his health have impaired his teaching. In spite of his preference for the difficult problems in mathematical physics, he knew how to enable his listeners to savor their studies which at the start could appear to be above the expected level of teaching. . . . His teaching is both profound and brilliant. This professor is absolutely devoted to his students, and in order to make himself useful to them, he never recoils from any additional work.[3]

Couat then added, on a puzzled tone, that in spite of his extraordinary performance as a teacher and a savant, Duhem, unlike even younger members of the Faculty, failed to obtain a salary increase. If anyone it was Couat, soon the head of a bureau of the Ministry of Public Instruction, who could have easily resolved his own puzzlement. Berthelot's shadow reached over the Ministry of Finances as well. Twenty-five years later, and already a member of the Académie des Sciences, Duhem still had to be satisfied with less than the highest salary available to anyone who had occupied a chair for ten or so years.

In Lille his first recompense was that he found there a brilliant group of students. According to his own recollections he owed it in good part to the insistent questioning of those students that in a few years he formed for himself a thoroughly reasoned idea about the nature, method, and competence of physics. Duhem, as he later recalled, entered the Ecole Normale as one convinced about the truth of mechanistic physics. Although at the Ecole he had Bertin, an unabashed sceptic about physical reality, for one of his teachers of physics, he left the Ecole still as a convinced mechanist. After arriving in Lille he quickly found out that it was one thing to study physics, and another to teach it, let alone to teach it to students whom only the full rigor of logic could satisfy. This was precisely the prompting needed by Duhem, the born theoretician. Within three years he started publishing a series of long essays on the method of physics, on theoretical and experimental physics, on the relation of physics and metaphysics. Other essays of this type of his were on the status of the ether (the bedrock of mechanistic physics) and on the different characteristics of physics as cultivated by Anglo-Saxon and Continental (French and German) physicists.[4]

All these essays were published in the *Revue des Questions Scientifiques*, a quarterly founded and published in Bruxelles by Father Carbonelle, a Jesuit. It served as the organ of the Société Scientifique de Bruxelles, an association of Catholic scientists, who counted among them some twenty members of the Académie des Sciences of Paris where most Catholic scientists did not dare to appear as a group. Word about Duhem's teaching must have quickly spread among Catholic intellectuals in Lille. They found it to be of great value for undermining a principal tactic of the anticlericals always busy spreading and perpetuating the slogan that science and faith are in irreconcilable opposition. They were the ones who called him to the attention of Paul Mansion, the noted mathematician at the University of Gand who in 1891 took over the editorship of the *Revue*. It was he who asked Duhem to send articles to the *Revue*. Obviously, he did not ask for technical articles, but for the kind which would reflect his classroom discussions of the nature and method of physics.

Those who spread the word about Duhem's teaching were those at the "Catho" who were also members of the Société Scientifique de Bruxelles. Some of them quickly befriended Duhem, who in turn helped them as best as he could. One of them was Monnet, an instructor of chemistry at the "Catho," whom Duhem welcomed in his classes and in the laboratory work he supervised. In doing so Duhem overstepped unwritten guidelines which imposed a rigid separation between the University and the "Catho," guidelines reflecting the senseless extent to which many at that time in France wanted to carry the principle of separation between State and Church. Nothing was more natural for Duhem than to help that very Monnet who in the spring of 1890 had introduced him to his future wife.

It was less an introduction than a carefully planned scheme. Monnet might have been prompted by Duhem's mother and sister, both of whom were concerned about Pierre's exclusive devotion to the cause of physics. Numerous and lengthy as were the philosophical and historical essays he published during his years in Lille, they were a mere trickle in comparison to what he published in physics proper during those same years. In 1891 there appeared the two volumes of his *Hydrodynamique, élasticité, acoustique*. By 1892 he published the three volumes, amounting to over 1500 pages, of his *Leçons sur l'éléctricité et le magnétisme*. A year later followed his *Introduction à la mécanique chimique*. The list of his publications for 1888-1893 contains over 40 scientific articles, in witness of the fact that he was regularly at his writing desk shortly after five in the morning each and every day. To his mother's and sister's suggestions about marriage he kept replying that "all is to be kept for science, there should be nothing between science and him."[5] Finally, Monnet succeeded in setting up the proper trap, a visit by the five Chayet sisters in the home of their uncle in Lille, Dr. Ernest Balthus, professor of medicine at the "Catho."

Within half a year after their first meeting in the spring of 1890, Pierre Duhem and Marie-Adèle Chayet, the youngest of the five sisters, were married. The wedding took place on October 28 in St Sulpice in Paris, a ceremony with deep spiritual significance for both bride and bridegroom. They religiously kept the manuscript of the sermon preached by the officiating priest. Soon they were able to implement the exhortation in a signal way. During their honeymoon, which took them to the Flemish seacoast, they met a seminarian. On learning from him that his education to the priesthood was in jeopardy because of financial problems, they decided to give the money which Pierre gave as a wedding gift to Marie-Adèle, or Maddie as she was called in the family, to that seminarian.

Obviously both had a trust in God's providence which had special designs for them. While the birth of their first child, Hélène, in September 1891, had not been without some complications, the second pregnancy was recognized already in its fifth month as posing a most serious threat to the mother. Young Madame Duhem carried on heroically. A week or so before the birth (it ended in the stillbirth of a son, whom the father baptized as he forced back his tears), Mme Duhem knew that her chances for survival were slim. Her last wish was that Pierre should remarry soon. On July 25, 1892, Pierre Duhem buried his wife and his stillborn son with her. He carried the blow for the remaining twenty-four years of his life with manly reserve, dignity, and faith. The memory of his wife was too sacred for him to think about remarriage.

Keeping his wife's memory so sacred entailed a way of life that demanded a constant reliance on God's grace. Duhem was already dead when a priest who knew him well recalled in public that Duhem frequently went to communion, a point invariably overlooked by those Duhem-specialists, invariably non-Christian or only nominally Christian, who expect to discover a "second woman" in his life. They should rather try to discover Christ and the way of life He can inspire and make possible. At any rate the tragedy of losing his wife foreshadowed the tragedy that was to be unfolding in full in his career.

The academic year 1892-93 became a particularly probing one for Duhem. The sense of deep personal loss weakened his inner resources on which his most conscientious teaching and feverish rate of publishing were a heavy drain. As one with a keen sense of duty, he could never suffer easily evidences of carelessness and insincerity. On being rebuked by his dean,

Demartres, for a biting remark about an assistant, he defended himself with a steely resolve, which so outraged the dean that he struck Duhem. Instead of the dean, who had previously sent glowing reports about Duhem to the Ministry in Paris, it was Duhem who was punished. In the summer of 1893 he was transferred to the University of Rennes, at that time easily the most somnolent provincial university in France.

Hurt Duhem must have been but surprised he would not be. By then it had become all too clear to him that Berthelot not only blocked his road to Paris but also made it understood by major publishers in Paris that they would incur his displeasure if they were to publish anything by Duhem that related to the principle of maximum work. A telling evidence of this is Duhem's letter, written from Lille on January 16, 1893, to a physical chemist who asked him to recommend a paper of his for publication in the *Journal de physique* with which Duhem had close connections.

> You undoubtedly don't forget that Mr Berthelot is almost omnipotent in France in matters relating to science and that it is not allowed to question there the principle of maximum work. And since the editor of the *Journal de physique* tells me that they cannot publish your paper, I really do not know where to turn. One cannot find in France a scientific journal independent of Berthelot. You will have some idea of this if I tell you that no publisher in Paris dared to take my *Introduction à la mécanique chimique* and that therefore I had to have it printed in Gand [in Belgium]. I am therefore forced, to my great regret, to send you back your manuscript. In any case, if I find the opportunity to make your ideas known, I shall rise to the occasion.[6]

In Rennes he found the University Library in an appalling condition for any serious scientific research. Even more appalling than this must have been for him the remark, "And now that Duhem is gone, what purpose will be served by the books he had ordered?"[7] — a remark made as he left Rennes a year later. Nothing short of appalling were the events behind his having to leave Rennes. In January 1894 he went to Paris, for the second time in a month, to ask Jules Tannery, for some time the Vice-Rector at the Ecole Normale, to help obtain a post for him in Paris. Duhem's trip followed a letter from Painlevé in which references were made to the apprehension of members of the Académie des Sciences, such as Darboux and Sarrau, both very sympathetic to Duhem's work, of Berthelot's turning against them were they to make a move on behalf of Duhem.

This second trip to Paris achieved just as little as the first. He returned to Rennes as the place from where he would never move again even if asked. Nine months later, to his great surprise, he received a telegram from the Ministry. He was to start teaching in early November at the University of Bordeaux. Duhem felt that he was used as a mere pawn on the academic chess board. Authorities in Paris, Liard in particular, sent him to Bordeaux without promoting him to the rank of professor, although he was to fill the post of a professor who had suddenly resigned.

At first he refused the transfer in obvious disregard of whatever damage he might do thereby to the remaining prospects, very slim, of his career. But then another telegram came from Paris, this time from Tannery, who conveyed Liard's words: "Tell your friend Duhem that he must accept, that he must understand that Bordeaux is the road to Paris."[8] Duhem took that telegram so much at face value that when his furniture arrived in Bordeaux, he told the movers not to unpack except what was absolutely necessary for daily living. It was not until months later that the two-storey house he rented at 18 Rue de la Teste, began to look as a home — his home for the rest of his life.

That Duhem was not sent as a professor to Bordeaux did not fail to trouble fair-minded people there. His arrival was announced with full awareness of his stature as a scholar in the *Rapport* for the year 1893-94 which went to press only a few days after word had reached Bordeaux about Duhem's appointment: "Mr Duhem comes to us with an established reputation, in full force and vigor."[9] Already on November 27, 1894, the Conseil d'Université unanimously asked the Ministry to set up a new chair for physics. In its message to the Ministry the Conseil called attention to the decay of the grand tradition of theoretical physics in France, initiated by Ampère, Fourier, Poisson, and Lamé. To recapture something of the foreign leadership, represented by Helmholtz, Thomson, Gibbs, and others, the Conseil suggested nothing less than an appropriate support for "Mr Duhem, one of the young and brilliant representatives of the school of Poisson."[10] On reading this, Liard could not fail to face up to the question: If Duhem was so brilliant as to offset singlehandedly the foreign leadership in theoretical physics, why was there no chair for him in Paris, the very place where France always wanted to compete with other nations? The setting up of chair for Duhem in Bordeaux on March 11, 1895, was a half-hearted move, shown by the fact that he received the lowest professorial rank and salary.[11]

The chief driving force behind that message of the Conseil was George Brunel, a noted mathematician and also the one who turned the dormant Société des Sciences Physiques et naturelles de Bordeaux into a first-rate scientific forum so that its *Procès verbaux* and *Mémoires* began to appear in all major libraries in Europe as well as in the United States. Duhem's contributions to both organs were frequent and at times book-length. Apart from those publications, he published between 1897 and 1899 his four-volume *Traité élémentaire de mécanique chimique fondée sur la thermodynamique*. An outgrowth of this work was a one-volume textbook that was immediately translated into English. Among his serial contributions not gathered into a single book were his essays on permanent deformation and thermodynamics that appeared in German in the *Zeitschrift für Physikalische Chemie*. The *Journal de mathématiques pures et appliquées* received from him in 1897 two articles on fluid mechanics, while the Académie Royale de Belgique was the beneficiary of his great memoirs on permanent deformation and hysteresis. In 1897 he began to send a series of articles to the *Journal of Physical Chemistry*, published at Cornell University, an illustration of the eager welcome of his ideas in the United States.

When on July 30, 1900, he was elected corresponding member of the Académie des Sciences, J. E. Trevor, editor of the *Journal of Physical Chemistry*, congratulated him with a stab at French officialdom: "I was greatly pleased to learn of this mark of official recognition which has taken so long to come." Professor Morrin, an older friend of Duhem from Rennes, offered a masterpiece of irony: "It is rather the Académie des Sciences that ought to be congratulated for having understood the impossibility of any further delay and for having given evidence, through a rare unanimity, of its regret of not having done earlier what it has just done today."[12]

Duhem's election, combined with the fact that he had received 36 of the 38 votes cast, should have become the cause for jubilation at the University of Bordeaux. Actually, everything was done by University officials to turn it into a non-event. In the *Rapport* for 1899-1900, reference to Duhem's election covered less than two lines, a circumstance that could not be accidental. Indeed, by 1900 the reports sent by the Rector about Duhem to the Ministry markedly differed in tone from the ones that characterized his first three years in Bordeaux. In all evidence, in Georges Bizos, who became rector following the death of Couat, some powers in Paris found a ready tool for their purposes.

From the very start Bizos took issue with Brunel, the dean of the science faculty, whose sole criticism of Duhem consisted in noting that he was not suited for administrative posts. According to Bizos, Duhem was rudely unfair toward those students "who do not belong to the Catholic and antirepublican côterie of the Ozanam circles. Imbued with clerical ideas in an extremly fanatical fashion, Mr Duhem is above all an ultramontane militant of the most violent kind."[13] Clearly, Bizos had an axe̅ to grind and seized on every opportunity to denounce Duhem in Paris. He in fact wrote a special letter to the Ministry after there appeared in the June 28, 1899, issue of the *Nouvelliste*, a Bordeaux daily, a speech which Duhem had delivered three days earlier at the meeting of the alumni of the Institut de Sainte Marie,[14] a *collège* run in Bordeaux by the same Marianists who were in charge of Collège Stanislas in Paris.

The speech which is reprinted in this volume as a faithful mirror of the unity in which Duhem saw Catholic religion and France tied together, was denounced by Bizos as a rank example of sinistrous agitation by Duhem against the highest officials of the République and against the République itself. What worried Bizos was that too many Catholics in Bordeaux refused to let their children be exposed to the atmosphere of the "laic" education. They knew more than enough about that virulently antireligious indoctrination in the name of *liberté, égalité, fraternité*. It must have simply infuriated Bizos that in Duhem he found a Catholic academic who did not hide his deepest religious convictions, in spite of the fact that this could mean, and in Duhem's case it certainly did, the turning of one's academic career into a protracted agony. Few other Catholics were willing to face a similar experience. Otherwise, in speaking about their student years at the Sorbonne, Gilson and Maritain would not have recalled their surprise on finding that this or that famous professor was a Catholic.

For the next five or six years, Bizos kept fulminating against Duhem, without noting the irony. In his confidential yearly reports to the Ministry, he described Duhem as a "perpetual firebrand who carries war everywhere he goes," and "who breathes constantly division and discord." In the official reports, the only ones he was obligated to show to Duhem, the latter was "an incomparable teacher most devoted to his task." Padé, who replaced Bizos in 1905 as rector, sounded from the start as one on the defensive. While he still spoke of Duhem's "bad character," he added almost apologetically: "Men of such caliber must be taken as they are."[15]

Though not a scientist, Padé could easily learn that by 1905 Berthelot and his scientific coterie had long since admitted defeat. A monumental proof of this came through the gala affair held at the Sorbonne on November 24, 1901, to celebrate the 50th anniversary of the beginning of Berthelot's scientific career. Only some sharp eyes, trained in thermodynamics, could see a dark shadow hanging over the glow of festivity. As speech after speech was delivered in praise of Berthelot, not a single reference could be heard to Berthelot's favorite brainchild, the principle of maximum work. Particularly telling was the silence on the subject in the longest speech of the day, a speech by Moissan, a future Nobel-laureate in chemistry, who dealt with the technicalities of Berthelot's chemical researches. If anyone it was Moissan, professor of chemistry at the Sorbonne, who should have risen to the defense of the maximum work principle if anything could be salvaged from it. The silence of Moissan and of others constituted an all too loud evidence that the young Normalien was right when the Sorbonne had judged him wrong.

That by 1901 official French science had to jettison the maximum work principle of Berthelot was due to the fact that the principle had been torpedoed in 1897 in a long article, much of it within the ken of non-chemists, which Duhem published in the *Revue des questions scientifiques*. The article was a review of Berthelot's two-volume *Thermochimie* of about 1600 pages. The work, mostly data of measurements, was received with muffled applause. Only Duhem dared to spell out the truth. He was all the more justified in doing so because just about when Berthelot published his *Thermochimie* he also read before the Académie des Sciences a short paper in which he made Duhem appear an upstart ignoramus.

Duhem knew how to handle one who did not dare to come out into the open by mentioning him by name. Through the *Revue des questions scientifiques*, which had in France and abroad thousands of Catholic subscribers, and as many if not more surreptitious anti-Catholic readers, Duhem carried the issue to the four corners of the scientific world. His chief point Duhem set forth by quoting a physicist most familiar with Berthelot's "proofs" of the maximum work principle. "He told me once," Duhem wrote, "that with such reasoning [as the one used by Berthelot] one can prove anything one wants to."[16] Another point consisted in laying bare once more the fact that all of Berthelot's ideas on the maximum work principle were an unacknowledged borrowing (plain plagiarism one should say) from the writings which Julius Thomsen, a Danish chemist, published in the 1860s. For vindicating his priority, Thomsen sent his autographed photo to Duhem.

What the broader public least could relish was for Duhem the most important matter, namely, the sacred cause of scientific truth. He believed in scientific truth because he was the kind of Catholic for whom truth, in Christ's very words, offered the only means of liberation in the deepest and broadest sense of this nowadays much abused word.

While defeated in the scientific arena, Berthelot refused to capitulate in the arena of academic politics. The consequences for Duhem's career were most painful. The chief of them was that for all practical purposes he was left with no graduate students, once the half a dozen or so students he had taught in Lille had followed him from there to Bordeaux and earned their doctorates under his guidance. In foreseeing himself giving his advanced courses before one or two students, he took the pen and reminded Liard in Paris of his words to Tannéry that "Bordeaux is the road to Paris." He recalled his reluctance to press his candidacy for any chair in Paris until all those who were his seniors had obtained one. He listed the eight doctoral theses defended under his guidance in merely five years, the honorary degree he had received from the University of Cracow on the occasion of its 500th anniversary, his having being elected foreign associate of the Société Hollandaise des Sciences, and his election as *membre correspondant* of the Académie des Sciences.

Then he turned to the actual situation in Bordeaux and to the immediate and long-range prospects there for teaching that theoretical physics which he had been sent to teach there in the first place. His chief complaint had for its target the Conseil d'Université which with its vote demoralized laboratory chiefs, able to do research, by denying them faculty status in spite of the fact that they had doctorates and gave courses as well.

> This vote, dictated by selfish considerations, has made to overflow the disgust which the actual state of the Faculty of Sciences in Bordeaux provokes in me. That situation made me decide to turn to you and ask you to provide me with a theater where my activity may produce some useful effect before that activity is killed by discouragement.[17]

A theater, that is, a chair for theoretical physics could have been easily found in Paris, or funded, if necessary. But in either case the call of Duhem to Paris would have been an admission of a signal injustice done to him in 1896 and 1898. In 1896 a beginner physicist, Gustave Robin, was asked to give the first course in physical chemistry at the Sorbonne, obviously because he was the son of an anticlerical senator who in addition also taught biology at the Sorbonne. Two years later, following Robin's death, Jean Perrin, still a beginner in the field compared with Duhem and nine years his junior, was asked to continue the course. After receiving Duhem's warm congratulations Perrin sent a reply in which plain dishonesty lurked between lines of obsequious courtesy. While holding high the value of Duhem's good wishes, Perrin expressed hope to see Duhem at the World Congress of Physics in 1900, still two years away. Then Perrin gave himself away by adding that he was going to Berlin and Leipzig to see there the prominent chemists, Van't Hoff and Ostwald. Why could Perrin not take the train of Bordeaux and consult with Duhem if the latter's good wishes were so valuable?

Perrin could be but apprehensive about a meeting with Duhem. News about it would have disgraced him in Berthelot's eyes. Furthermore, he knew that by then he had become guilty of rank plagiarism of Duhem's ideas through the paper he had presented on the laws of thermodynamics before the Société Française de Philosophie. He also knew that he was to succeed Robin whose course constituted a ransacking of various books of Duhem. This became public knowledge, carefully ignored by officialdom, when the text of Robin's course was published posthumously in 1901.

Even a Poincaré made furtive use of Duhem's ideas, but he was at least rebuked on the spot by Hadamard. Hadamard's esteem for Duhem must have been very deep, otherwise he would not have risked a confrontation with Poincaré, possibly the greatest name in mathematical physics around 1900. But Hadamard was also a great figure in pure mathematics and, furthermore, he did not have to fear losing his chair at the Sorbonne or his membership in the Académie des Sciences. The point at issue was the example which Poincaré used in the lecture he had given at the World Congress of Physics about the revisability of physical theories. The example showed the enormous conceptual complexity of any step in the measuring of electrical phenomena, which in turn illustrated the difficulty of establishing one-to-one correspondence between physical reality and theoretical constructs about it, let alone establishing those constructs as absolutely definitive. Hadamard pointedly recalled that the very same example had already figured prominently in a long essay of Duhem on theoretical physics published in 1893.[18]

Most of the time Duhem left without comment the many
forms of plagiarism of which his writings were (and still are!) a
constant target. Having his mother live in his home, it must
have been easy for him to recall her readiness to help him with
her prayers and sympathy when toward the end of his education at Stanislas he had first encountered that painful facet of
academic life. If anyone, then his mother knew how deep was
his sense of justice and how conscientiously he acknowledged
in his writings any idea or information he owed to others. Her
tactful reminders about the need for patience and forgiveness
enabled Duhem, as years went by, to smile on finding ever new
instance of his writings being ransacked by fellow academics.
His daughter, only ten or so, vividly remembered the moment
when, sometime in 1901, he said to his mother: "I no longer
count these robberies. I would waste my time by reclaiming
from the Académie des Sciences a recognition of my priority.
Let them pillage me as much as they wish, if this pleases them;
I have plenty of ideas for all of them; physics will lose nothing!"[19]

The measure of comfort he was able to derive from his
dedication to the cause of physics — his lay vocation — was
too deep to be fathomed adequately. While during 1901–1904,
evidently the bleakest period in his academic career, he needed
that comfort, it was for something more concrete, a little book
in the drawer of his writing desk, that he reached out time and
again for strength. It was his copy, worn through frequent use,
of *The Imitation of Christ*. He knew it almost all by heart.
Appropriate passages from it came to him naturally. A most
memorable case of this is in his devastating exposure of
Berthelot's inane defense of the maximum work principle,
which he brought to a close with a portrayal of the gradual
dying out of the voluble praises once accorded to it: "Tell me,
where are now the great students and famous scholars whom
you have known? When alive, they flourished greatly in their
learning, but now, others have succeeded to their posts and

promotions, and I cannot tell whether their successors give them a thought. In their lifetime they were considered great in the world, now, little is spoken of them."[20]

About Duhem, tried so heavily in his patience and readiness to forgive, exactly the opposite was to become true. While Berthelot's writings gather the dust, and the same is true of those many who by ransacking Duhem's writings earned much false credit in their lifetime, Duhem's books keep being reprinted and translated, and are the object of many articles, dissertations and monographs. He was fully satisfied with that long-range   prospect of proper recognition. His faith in God and recompense in eternal life gave him the peace of mind which he needed to go on with his exacting intellectual efforts and to cope with two burdens immediately ahead of him, of which one,   relating to his daughter, became a protracted source of deep concern, a sort of tragedy for him.

In the opening years of the 20th century Duhem's chief personal concerns related to the future of Hélène, who had just turned ten. His hope that his sister Marie would eventually take over from his mother Hélène's education was shattered when Marie entered religious life in 1898. His mother was visibly weakening. As to his own health, it was not commensurate with his enormous will power. Each spring and summer he suffered enormously of hay fever. Some mild heart attacks he simply ignored. Nor did he give proper attention to the rheumatism of the stomach he had contracted as a boy. He kept asking God to keep him alive only until Hélène had safely reached adulthood.

In those bleak years of 1901-04, one of the comforts he received from above was the arrival in Bordeaux of a young professor of Church history, Albert Dufourcq. Disregarding the rumors that Duhem was inaccessible, Dufourcq paid his respects at 18 Rue de la Teste. Before long, Duhem was a weekly guest of the Dufourcqs and a great favorite with the Dufourcq children who were to come in quick succession. It was through that friendship that Providence assured, many years later, a much needed protection for Hélène against a rude exploitation, an intellectual guidance in her great lifework, and a most generous assistance in her last years.

Another providential turn in Duhem's career came when he discovered, in 1904, Jordanus and, a year or two later, Buridan, and Oresme. They came within his purview at a critical moment as so many unsuspected witnesses on behalf of his search for the perfection of physics, a search which had a historical perspective. Hélène herself recalled the enthusiasm with which her papa used to speak in the evenings at home about those mysterious medievals. Work once more proved to be a most efficient antidote to ever new cases of slighting which Duhem keenly felt also when his friends were the victims. One of them was Paul Tannery, the great historian of Greek mathematics and astronomy, who did his research while he had to earn his living as a supervisor of a state-owned tobacco factory. Instead of Tannery, a second-rate crystallographer obtained the chair for the history of science in the Collège de France when it became vacant in 1903. The injustice, that made headlines, proved to be too much for Tannery. He died shortly after he had presided at the Congrès International d'Histoire des Sciences in Geneva in 1904. In his obituary of Tannery, Duhem asked the question which only someone fearless of consequences for his career, but also fearlessly devoted to moral principles, dared to ask:

> An injustice produces at times grave though remote consequences. The one who did the injustice will have to answer for the remotest results. Is this principle of ethics kept in mind by those who place in the service of sects and parties the power with which they have been entrusted for the sake of the public good?[21]

The sects Duhem had in mind was secularism in general, and Freemasónry in particular. For some time prior to Duhem's times those sects had clearly perceived the usefulness of creating an image of the origin and history of science that would show Christianity as a fundamentally antiscientific movement. No wonder that when first established in 1893, the chair went to Lafitte, who earned far more credit as the "pope" of the positivist church than as a historian of science. As a diligent Freemason, Wyrouboff, the crystallographer in question, was a perfect choice to succeed Lafitte.

　　Duhem's indignation about the injustice done to Tannery for patently ideological reasons had to be all the deeper, because he knew two things: One was that Tannery had already exposed the fallacy according to which the decline of Greek science was due to the rise and spread of Christianity. The other related to the discovery he was just making around 1905 about the medieval or rather Christian origins of Newtonian science. As has already been noted, the first beneficiaries of that discovery of his were his mother and daughter with whom he kept sharing his excitement. He was not to enjoy too long that most appreciative audience.

　　A year or so after Duhem's mother died in 1906, his daughter began to spend more and more time with her aunt, Marie Duhem, living in Paris as a nun expelled from her convent. Before long Hélène took up permanent residence with a group of women dedicated to Catholic social work just outside Paris. From 1909 on father and daughter spent together only the second part of the summer and the first month of the fall in Cabrespine. While in Paris, Hélène received daily letters from her father who only on occasion let her feel something of the burden of his solitude.

　　Duhem's solitude was, however, diminishing as far as officialdom was concerned. A telltale sign of this was a remark in the report of Padé, who replaced Bizos as rector in 1905, that at the university "a sort of veneration surrounds Duhem." A year later, all earlier accusations against Duhem as a man of bad character were unmasked in Padé's report: "As a man he is apt to be indignant. But I have never seen him cause any difficulty except when he believed that common interest was being violated."[22]

It was the unselfishness of Duhem which prevented that the many honors coming his way did not, as often happens in academic circles, produce an air of jealousy about him, but rather an aura of sincere appreciation among his colleagues. In 1902 he became foreign associate of the Académie Royale de Belgique. In 1905 he was elected member of the Polish Academy of Science in Cracow. In 1907 he received the Petit d'Ormoy Prize from the Académie des Sciences in Paris, a prize worth 10,000 francs, or 2000 francs more than his annual salary between 1904 and 1910. In 1909 he was elected corresponding member of the Dutch Society of Experimental Physics of Rotterdam and in 1912 of the Reale Istituto Veneto di Scienze, Lettere e Arte. Soon afterward the Reale Academia di Scienze of Padua made him honorary associate.

All these honors, however gratifying, did not affect him much. To the Secretary of the University of Bordeaux, who in 1909 asked him for a list of such honors, he sent it with the note: "Please, file this list with a view to my future obituary notice."[23] This was not affectation of humility on Duhem's part. His mind, far from being riveted on honor, was being drawn ever more deeply into the exploration of the field he had just discovered, the vast tracts of medieval science. His career at the University of Bordeaux quietly shifted into that of a lecturer to all the Faculties.

First came in 1904 and 1905 a series of lectures on the philosophy of science. They were based on the text of his great classic, *La théorie physique: son objet et sa structure* (*The Aim and Structure of Physical Theory*), to be published in 1906. Those lectures were followed from 1906 on by a series of yearly lectures on the history of cosmological doctrines from Plato to Copernicus. Their text eventually formed the first five volumes of his *Système du monde*, undoubtedly the most original and monumental scholarly work produced in this century by one who had to do, with practically no help from others, all his research and writing.

Duhem's mind was set on a goal transcending time. He saw his immense labors "sub specie aeternitatis." To friends who felt that several lifetimes were needed to bring his project to completion, he kept saying that if it was useful in God's eyes, He would not let it go unfinished. He practically finished it in ten short years, though at a price. As one whose favorite motto was that hard work had not yet killed anyone, he must have felt enormous satisfaction on reading the rector's report in 1914 in which it was mentioned that "Duhem works often to the exhaustion of his strength."[24]

By then it was public knowledge that in May 1913 Duhem had assumed a superhuman obligation by signing a most extraordinary contract with the publisher, Adolphe Hermann in Paris. Beginning with 1913 he was so send to Hermann each year, for the next ten years, a manuscript equivalent to 500 printed pages, while the publisher obligated himself to bring out a volume of that size each year. Duhem waved royalties for the first 400 copies sold of each volume, and was assured 40 percent of the royalties for any further copy sold.

Duhem's dedication to the common good, his disinterestedness in honors, and his dedication to exacting scholarly work, whose rewards he was hardly to reap personally, bespeak of an unusual measure of unselfishness, possibly the finest test of one's Christian convictions. His sincerity could not be doutbed as he expressed little interest in his being elected as one of the first six non-resident members of the Académie des Sciences. Strong words from his former teacher, Darboux, who told him that as a member of the Académie he could do more good, were needed to convince him that he must submit his candidacy. Even after that he authorized Edmond Perrier, a member of the Académie des Sciences, to make public his readiness to yield his place to Henri Fabre, the famed and very old entomologist, who was left off the list of candidates.

While others became increasingly agitated as his election was being postponed over six months, he remained supremely undisturbed. He knew that those in Berthelot's coterie did everything to prevent his election into the Académie. Although he should have been elected already in July as the second of the six, he became the last of the six on December 2, 1913. He derived less pleasure from his election than from the fact that the Ministry of Public Instruction decided, earlier that year, to buy 300 copies of each of all the volumes of the *Système du monde*, partly for distribution among French libraries. He was just as overjoyed when about the same time Darboux informed him about the release of 2000 francs from the Debroux-funds to support the publication of the first volume of the *Système du monde*.

There is little evidence that Duhem in 1913 had any premonition about the very few years that remained to him. His eyes, ever fixed on lasting principles and prospects, saw through the blaze of glory that suddenly surrounded him. Two months before his election, he wrote his daughter:

> I have just read your letter of yesterday. It seems to me that you entertain plenty of illusions concerning the impor-tance which that title of "Membre de l'Institut" will have for me. I will be asked, a little more frequently than in the past, to preside over committees and assemblies — things which I abhor, but my books will not be more widely read, my ideas will not be more often discussed, the only thing which I desire. You tell me that I have more influ-ence since I am a corresponding member of the Institut; I believe the opposite is true; my works have passed more and more unnoticed. This year, one copy was purchased of my great treatise on electricity. For me that title will have the effect of a crown deposited on the coffin into which fellow-physicists have nailed me alive.[25]

What he wrote was a prophecy in more than one sense. Following his election he was more sought out, but not much more. He was also to live the remaining three years of his life as one practically dead as far as Perrin, Langevin, Mme Curie and other French protagonists of atomism were concerned. Their names cannot be found on the hundreds of letters and telegrams of congratulations Duhem received in December 1913. For them Duhem remained a dinosaur from whose influence the younger generation of French physicists was to be protected.

The Nobel-laureate physicist, Kaestler, who served briefly on the physics faculty in Bordeaux in the late 1920s, recalled fifty years later that at that time some younger physicists there spoke with resentful disparagement of Duhem the physicist. Whether Duhem took the wrong tack, a tack riveted on the past, or whether he at best froze into that present which goes by quickly, or whether the tack he took pointed really to the future, are questions to which we must now turn. Proper answer to them is essential to an understanding of the true contribution of Duhem to a reliable perspective about the harmony between faith and science.

4

The Perfection of Physics

It is only natural that an author should intensely want his books to be widely read. In mid–1913, when Duhem expressed this very desire to his daughter, he must have known that his book, *La théorie physique: son objet et sa structure*, was on its way to become a classic. It had already been translated into German and, within a year, it was to come out in a second edition, enlarged with his impassioned reply to Rey's charge that his physics was the physics of a believer. By then Duhem was about to receive the galleys of the first volume of his *Système du monde*. He must have felt confident about that volume and the volumes to follow that they would reach many readers.

Had he ever considered himself a philosopher or a historian of science, he would not have written his daughter that what he really wished was not his election to the Académie des Sciences but that his books be widely read and discussed. He meant his books on physics. From his years in the preparatory school of Stanislas, a physicist he wanted to be and throughout the rest of his life he wanted to be known as a physicist and nothing else. When in 1904 the chair for the history of science in the Collège de France became vacant, some people thought that Duhem might be interested. His daughter keenly remembered his reaction: "I am a theoretical physicist; either I shall teach theoretical physics in Paris, or I will not return there."[1]

Long before 1904 Duhem had every right to claim a chair in Paris. His formulation of the thermodynamic potential, although not acknowledged in France, was widely hailed in that very Germany which loomed large as the supreme standard of erudition in the eyes of ever larger numbers of French intellectuals. Wilhelm Ostwald, easily the leading chemist of the time, spoke in superlatives of Duhem's massive commentaries on the principles of thermodynamics, published in three parts in 1893–95. Ostwald greeted the *Potentiel thermodynamique* when republished in 1896, as a "work that played an important and influential role in the rapid development of the application of thermodynamics to the phenomena connected with physical and chemical equilibria."[2] The fourth volume of Duhem's *Mécanique chimique* appeared to Ostwald "as another monument to the famed author's brilliant methods."[3]

Compared with the encomiums coming from Germany, significant appreciation of Duhem the physicist rarely appeared in print in France in the 1890s. In one instance, the appreciation revealed a studied reluctance to give Duhem his due. "Twice in the book," Poincaré concluded in the 1892 introduction to his lectures on thermodynamics, "I happened to be in disagreement with Mr. Duhem. He might wonder that I cite him only to combat him. I would be saddened if he thought of any ill will. I hope he will not suppose that I ignore the services he had rendered to science. I have only thought to be more useful by insisting on points where his results seemed to me to be in need of being complemented rather than on those points where I could but repeat him."[4]

Clearly, if Duhem's services to a major branch of science were so important that even a Poincaré could but repeat him, he should have deserved something other than two critical remarks. Moreover, very important were declared about that time Duhem's contributions to another branch of physics, electromagnetic theory. In 1893 Painlevé, already professor of theoretical mechanics at the Sorbonne, characterized Duhem's three-volume work on electricity as a continuation of the researches whereby Duhem had already given a new theoretical unity to hydrodynamics, viscosity, vaporization, and dissociation. Against the background of "innumerable treatises of very unequal merit" that since Poisson had appeared on the subject, Painlevé described those three volumes as the kind of "daring enterprise which Duhem is not afraid to undertake." Few physicists at that time, or at any other time, could expect a better encomium than the one which Duhem received from Painlevé: "The *invariably analytical* method adopted by Duhem gives to his book truly the character of *power* and *unity*" (italics added.)[5]

While the first printed appearance of the expression, "Gibbs-Duhem equation," postdated Duhem's death by half a dozen years, physical chemists had long before that seen its fundamental importance. There is a similar story to the Duhem-Margules equation. It took more than half a century and a first-rate expert, Truesdell, on theoretical mechanics before there should appear in print the expression, "Clausius-Duhem inequality" and its appraisal as a "pillar of the mathematical theory of thermodynamics."[6] Yet the advantages that could be derived from those equations by experimentalists were not such as to result in startling discoveries. That Duhem gave the first rigorous definition of reversible transformations did not necessarily give practical help even to those specializing in thermodynamics. Furthermore, following Becquerel's and Roentgen's discoveries, the new and exciting fields of radioactivity and X-rays drew the attention of physicists even more strongly toward experimentation. Such an outcome hardly favored even in France a revitalization of a once great interest in theoretical questions.

Not that Duhem ignored these new fields of discoveries. A telling proof of this came from George H. Bryan, leading

British thermodynamicist around the turn of the century and eventually the president of the London Mathematical Society. As he visited Duhem in Bordeaux in 1901 he was very much impressed by the way Duhem taught experimental radioactivity. Duhem's single apparatus appeared to him far superior to the elaborate but "dusty instruments" he had earlier seen in Boltzmann's laboratory in Vienna. Bryan went, of course, to Bordeaux, to meet Duhem, "a shortish man with a very pleasing manner," the theoretician bent on offering as vàst a system as possible. Bryant correctly saw the true nature of Duhem's work, "a large portion of which is highly original," and he saw it precisely in that light which explains why Duhem "did not concentrate his main efforts on the discovery of new phenomena or the measurement and remeasurement of physical constants." Duhem played, Bryant continued his appraisal of Duhem's work, "at least an equally important part in the advancement of our knowledge by evolving order out of chaos, and uniting isolated portions of mathematical physics in the form of a connected and logical theory."[7]

Logic is, however, useful only inasmuch as it helps one see order in the welter of real facts. Toward the end of his life Duhem had to recognize that facts did not bear out his sustained efforts to supplant Maxwell's electromagnetics with that of Helmholtz. Long before that he had to hear from none other than Pierre Curie that his warnings against the logical shortcomings of Maxwell's theory sounded rather hollow. "I think," Curie wrote in 1902 to Duhem, "it would be a good idea that our physicists display in Maxwell's style an *unheard imprudence*," the very thing Duhem decried. Curie did so in acknowledging a courtesy-copy of Duhem's freshly printed *Les théories électriques de J. Clerk Maxwell*, a more than 200-page-long critical and historical study, and asked: "By what to replace Maxwell's way of reasoning?" and voiced his repugnance "to go back to purely mathematical formulas which represented nothing physically."[8]

Curie grossly exaggerated the case. Duhem, who had by then for years kept insisting that in an ideal or perfect physical theory all mathematical details should relate to aspects of physical reality, would not have contradicted himself to the extent suggested by Curie. Duhem's theory had in fact for its central contention that there should be longitudinal electromagnetic waves, an inference that could very much challenge the experimental investigations. Duhem was, of course, wrong when a few years later he greeted Blondlot's experiments as a demonstration of those longitudinal waves, inadmissible in Maxwell's theory.

The usefulness of theoretical work for experimental research may take much time to show through. Only recently has it been recognized that Duhem's work on viscosity contains parts, such as the one relating to eddies that do not slide past one another, that are very valuable in the study of plasma which is central in fusion research. But in Duhem's case the search for new facts slowed down precisely at a time when new facts were pouring into the physicist's ken along a very broad front. The two volumes of his *Traité d'énergétique*, published in 1911, his most memorable effort in theoretical synthesis and in a sense the synthesis of all he had achieved in theoretical physics, reflected a course he had given seven years earlier.

When Duhem published the *Traité d'énergétique*, which contained no discussion of spectral lines, atomic theory, quantum theory and relativity, these topics were already taking the center stage in leading periodicals and major conferences. The year which saw Duhem's election to the Académie des Sciences, saw also the publication of Perrin's *Les atomes* and its going through five editions. The chapter that in 1907 brought Duhem's book *On Absolute and Relative Motion* to a close had "A Glance at Modern Times" for its title. The "modern" material treated there began with Descartes, continued with Newton and Kant, and did not go beyond Carl Neumann and an early book of Ernst Mach. Neither there nor elsewhere did

Duhem ever discuss in a systematic way the relativity theories of Poincaré and Lorentz, let alone of Einstein.

Nothing should seem therefore more tempting than to write off the *Traité d'énergétique* as a rear-guard action. Yet, three-quarters of a century after its publication its riches remain far from being fully discovered. On February 23, 1981, in a letter to this author, Truesdell wrote: "Without ever having made a systematic study of Duhem's voluminous writings, we continue to find things there. For example, there is the *Traité d'énergetique*, a dense and forbidding work, but in it we found just a few years ago the basic idea of using a Liapounov function to relate dynamic stability to static stability in a deformable continuum." And as if to make the irony complete, Truesdell added: "This idea had recently been proposed as original by one of the most eminent men in our field."[9]

The *Traité d'énergétique*, almost a thousand pages of pure science and totally free of the pseudo-mysticism into which Ostwald steered his own brand of energetics, must appear a prophetic work precisely because of its author's antiatomism. Gone are the heady days when the discovery of the electron and of the atomic nucleus convinced many that the ultimate building blocks of matter had come thereby in full view. Einstein himself told his friend, Moszkowski, around 1920 that there was nothing to be discovered beyond those particles. The number of "fundamental" particles has since grown into an embarrassingly large quantity. Moreover, enough time has gone by to bring in full view the implicit absurdities of the Copenhagen pseudo-philosophy of quantum mechanics. It should be enough to think of the idea of a "passion-at-a-distance," which some supporters of that pseudophilosophy felt bound to read into the experimental verification of Bell's theorem. Physicists may, before long, find much-needed food for thought in a seldom quoted remark which Dirac made in 1979 at the Jerusalem Einstein conference:

It seems that at present quantum mechanics is not in its final form. Some further changes will be needed, just about as drastic as the changes which one had made in passing from Bohr's orbit to quantum mechanics. Some day a new relativistic quantum mechanics will have determinism in the way that Einstein wanted. This determinism will be introduced only at the expense of abandoning some other preconceptions which physicists now hold, and which it is not sensible to try to get at now; . . . so under these conditions I think it is very likely, or at any rate quite possible that in the long run Einstein will turn out to be correct even though for the time being physicists have to accept the Bohr probability interpretation — especially if they have examinations in front of them.[10]

Dirac meant, of course, Einstein's persistent insistence on that continuum which underlies the meaningfulness of classical differential equations and, by inference, the possibility, however purely theoretical, of perfectly accurate measurements. This is not to suggest that physicists certainly will again take the notion of continuum for the theoretical basis of their work. But in an even deeper sense, Duhem's principal contention seems to have already been vindicated by the actions if not by the reflections of leaders in fundamental particle physics. Already the very large number of such particles makes the label "fundamental" appear rather inappropriate. More importantly, they have long ago ceased to appear particles in the ordinary sense of that word.

To a greater degree than one may suspect, the physics of the 1970s and 1980s has become what Duhem, a century ago, claimed physics ought to be: a logically satisfying systematization of energy measurements. It is a secondary matter that energy on the atomic level and below it appears discontinuous. What Duhem would point out today is that, just as he claimed it a century ago, the mathematical formalisms of physics do not entitle the physicists to make absolutely final declarations even about the purely quantitative features of material entities, let alone about their very nature.

In facing the claim, for instance, that quarks or gluons are to be taken for absolutely indivisible units of matter, Duhem would merely repeat that "no reasoning can engage those unconcerned whether they are right or wrong." He said this toward the end of the long summary of his work as a physicist in the *Notice* which he submitted to the Académie des Sciences prior to his being elected as its member. There he also emphasized that the aim of physical theory "was to classify and to coordinate the chaos of acts revealed by experience."[11] Only such a method could logically support the "PERFECTION OF SCIENCE" which he felt was the true aim of the physicist's labors.

Those words in capital letters stood as such in an essay of his on physical theory published in 1893.[12] Clearly, he remained consistently devoted to a purpose which he had espoused early in his career. One consideration that kept supporting him against extraordinary odds came from his trust in human reason, and especially in its logical powers. That trust he needed all the more as toward the end of his life it was clear to him that his books on physics were not being read. In facing the apparent prospect that his life work was wasted, he took comfort from the endurance of logic. In the *Notice* his almost hundred-page-long survey of the work he had done as a physicist had this for its final note: "Eternal as it is, logic can afford to be patient."[13]

Had reason meant for Duhem but mere logic, he would have become a forerunner of logical positivists, some of whom tried to claim him. But while the logical positivists remained caught in their logicism, which deprived them of the right to know reality external to them, Duhem never despaired of knowing reality. From his Lille-essays on he held high the fundamental role whereby metaphysics alone assures for the physicist that reality whose quantitative aspects he is to ascertain and to correlate.

One can only regret that Duhem did not develop systematically the philosophically most pivotal among all his statements:

> When a sincere witness, sufficiently sober so as not to take the whims of his imagination for observation and familiar with the language he uses to express his thought clearly, affirms to have registered a fact, the fact is certain. If I declare to you that on such a day, at such an hour, in such a street of the city I saw a white horse, you must believe, unless you have reason to consider me a liar or a victim of hallucination that on that day, at that hour, in that street, there was a white horse.[14]

Such a statement, so relevant to the very foundations of a realist epistemological stance, could have easily been developed, as it was later in Gilson's hands, into the epistemological system of what Gilson felicitously called, "methodical realism."[15] Duhem made no effort to develop such a system. He even failed to add that such an immediate knowledge of reality is implied in any effort aimed at casting doubt on that very knowledge. Had he developed that statement of his to even a modest degree, Duhem might even have discovered the role which his Catholic faith played in keeping him a realist while he could "logically" be expected to drift into the positivism of Ernst Mach for whom all reality was ultimately reduced to mere sensations.

For in a much deeper sense than Duhem ever hinted at, his physics was that of a believer, or rather of his Catholic faith. That faith of his rested on two propositions, the acceptance of both of which meant a thorough commitment to realism. One proposition consisted in the very first tenet of the Creed, belief in the Omnipotent Father, Maker of Heaven and Earth, of all things visible and invisible. Such belief had for its basis the recognition of a real "all" or universe. Duhem did not think that science in general or physics in particular has propositions applicable to the universe as such. He explicitly rejected the applicability of the law of entropy to the universe. In that connection too he refused to follow the fashion among physicists, some of whom quite notable, such as Helmholtz and Kelvin, who conjured up in vivid prose the heat death in store for the universe.

Of course, Duhem died before Einstein published in 1917 the fifth instalment of his memoirs on General Relativity, the one dealing with its cosmological consequences. There Einstein showed that it was possible to deal with the totality of gravitational matter and not to incur at the same time the gravitational paradox of an infinite, homogeneously distributed mass, often taken for the Newtonian universe. In other words, Einstein reinstated from the scientific viewpoint, the intellectual respectability of the notion of the universe, the very target of that "critical" philosophy which Kant tried to sell in the name of science.[16] Catholics, sufficiently aware of the fundamental importance of the solidity of the cosmological argument within their credal system, still need to develop broad awareness of the support they can derive from the cosmology of General Relativity.

Duhem never dealt either with the gravitational paradox or with its optical counterpart, although they had been often discussed around the turn of the century. There is a puzzling touch to the concluding part of his "Physics of a Believer" where he states that

> if we rid the physics of Aristotle and of Scholasticism of the outworn and demoded scientific clothing covering it, and if we bring out in its vigorous and harmonious nakedness the living flesh of this cosmology, we would be struck by its resemblance to our modern physical theory; we recognize in these two doctrines two pictures of the same ontological order, distinct because they are each taken from a different point of view, but in no way discordant.[17]

What did he really mean by that close parallel between the cosmological thinking demanded by science and the cosmology championed by the Church? He could not of course mean that geocentric system which the Church, contrary to some claim, never made part of its official teaching. But he remained just as generic about scientific cosmology as he did about Scholastic cosmology. About the latter he spoke briefly in one of his Lille essays as a philosophical study of motion, taking this word in the sense of any transformation.   Such was certainly the meaning he gave to motion in his generalized physics, or *Energétique*.

Whatever else Duhem had in mind, it must have included something connected with the progress of science on which he had very specific views. As science progressed, its laws gradually approximated a classification of empirical data that corresponded to a natural state of affairs, that is, to the real nature of things themselves. Therein lay the most Aristotelian aspect of Duhem's natural philosophy which would have become a contradiction in terms if he had not held the reality of a Nature or universe as something existing independently of the observer. But here too he stopped short of articulating himself. The reason for this lay in his resolve — at times explicit, at times implicit — to do philosophy only inasmuch as it was demanded by the immediate needs of doing physics as well as possible.

In that connection, he felt it was enough to make brief though incisive statements such as the one already quoted about sighting a white horse. It was in fact not in print but in a letter to a friend that he made his most significant declaration about his life-calling: "I have held it my duty as a scientist as well as my duty as a Christian never to cease being the apostle of common sense, the sole foundation of all scientific, philosophical, and religious certainty." In that letter Duhem also faced up to the question whether the claims of common sense were not "tantamount to some philosophical and religious beliefs, all resting on worthless reasonings which invariably imply undefinable notions, so many empty words void of meaning." In answer Duhem noted that

> the same could be said in connection with all the sciences, including those which are considered the most rigorous among them — physics, mechanics, and even geometry. The foundations of any of these constructs are formed by notions, which one pretends to understand although one cannot define them, or are formed by principles which one feels assured about, although one has no proofs of them whatever. These notions, these principles are formed by common sense. Without this basis provided by common sense, a basis not at all scientific, no science can maintain itself; all of its solidity comes from there.[18]

A realist philosopher, appreciative of that reality which is metaphysical even in ordinary sense perceptions insofar as all these are understood as replete with universal meaning, would make two remarks about Duhem's endorsement of common sense. For one, he may find Duhem's separation of common sense from metaphysics to be unnecessary and perhaps even misleading. For another, he would wish that Duhem had been more articulate on the subject. But in all his public endorsements of *bon sens* as basic to all right reasoning he never probed into the various meanings of that expression, or its difference from its trivialized version, a mere *sens commun*, or common opinion. Duhem felt that common sense was clear enough, and certainly clear for his specific purpose, namely, the analysis of a physics resting on that common sense as its very foundation.

It was in analyzing the "superstructure" of the method of physics that he drew graphic and logically unanswerable details about its limitations, provisional character and on its dependence on the cultural (national) background of physicists. In that latter respect he was certainly a pioneer. To many of his contemporaries, especially in France that resounded with Renan's famous declaration that "science will organize God himself," nothing short of a shock was contained in Duhem's conclusions about the aim and structure of physical theory. Those conclusions meant that the most exact form of science, physics, cannot be seen as a set of propositions that had organized themselves, so to speak, into a definite system.

Too many in France and elsewhere took for their supreme teacher about science a Renan, who blithely published in 1892 a book, *Avenir de la science,* by then more than forty years old in manuscript. Although it was sufficiently known that Renan had not even a remotely adequate training in mathematics and physics, much credit was given to his claim made there that "mathematics and inductive physics have always been the basic elements of my thinking, the only stones of my intellectual masonry that have never changed place and always served me well."[19] To his contemporaries, keen on taking the "scientific" method for the only respectable form of rationality, Duhem's dissecting that method suggested that physics, this most exact form of science, cannot organize itself. Therefore Duhem's work could but appear a supreme threat, best to be ignored.

And ignored it was. Duhem was already dead for two years when Hélène Metzger, a doctoral candidate at the Sorbonne in the philosophy of science, experienced the shock of her intellectual career. It consisted in hearing from one of the jury examining her thesis on the conceptual development of crystallography, that its principal point had formed the subject of some most interesting pages of Duhem's *Théorie physique.* This is all the more telling because the Duhemien provenance of that point is not revealed in *The Structure of Scientific Revolutions* by Thomas S. Kuhn, who gave a generic credit only to Metzger,[20] although by 1961 the *Théorie physique* had been widely read in English translation. The point is that Duhem saw a chief limitation of the scientific method in the fact that it was confined in each age to what could be accommodated within a leading idea, which Kuhn called a paradigm.

Duhem would have been the last to infer from this that the sequence of those ideas constituted an illogical chain, destructive of the notion of scientific progress. Duhem would have been the last to yield to a nominalist or an idealist temptation to "reify" those notions and lose thereby a hold on reality which they were meant to designate. In fact, it was Metzger herself who expressed in her book, *Les concepts scientifiques* (1926) where she generously recognized Duhem's priority, full agreement with Duhem's "apparent nominalism under which there lies disguised a very deep realist conviction."[21] She did not live to see the sinister consequences of a systematic slighting of that conviction. It came in full view in the measure in which philosophers of science from the 1960s on succeeded in creating the impression that their field was philosophy itself. As they severed their ideas about science from reality, they turned scientific ideas into factors that did ultimately the scientific thinking that is supposed to be done by real scientists.

In that sense, much of the philosophy of science written during the last half a century is an abuse of Duhem's seminal ideas, an outcome helped by not a few Catholic philosophers of science who, lest they should offend their secularist colleagues, failed to cultivate Duhem's thought. At times they tried to outdo their secularist counterparts by fanning the winds of destructive academic fashions. There should seem something sinistrous in the fact that a prominent Catholic University in the United States gave an honorary degree to Thomas Kuhn whose *Scientific Revolutions* contradicts that natural theology which, according to Vatican I, is part and parcel of Catholic teaching.[22] There is more than what meets the eye in the fact that a few years later Thomas Kuhn went to Tübingen, at the invitation of Hans Küng, to teach there to an international gathering of younger Catholic theologians the art of how to relativize dogmas by casting this evolution into Kuhnian paradigms. Duhem saw an antidote to relativism in the slow evolution of ideas which also assured their coherence, a point he emphasized in various contexts.[23] It was within that restriction that Duhem endorsed the biological struggle for life as illustrative of the competition of ideas, scientific and other.[24]

Had Duhem been cultivated among Catholic philosophers of science, they would have found in his writings short but stark warnings about the importance of mental strategy. For Catholics the success of that strategy hinges on the resolve to vindicate the essentially metaphysical character of all basic notions used in physics, such as body, extension, time, motion (notions listed by Duhem). The success depends on not conceding, in any form whatsoever, the claim that a physicist can, by sheer reliance on the method of physics, justify ordinary existence statements on which all his work depends, namely that such and such an instrument is in front of him, or that such and such a physical process does indeed take place. The success depends on the resolve to allow no more to the method of physics than a competence to register and co-ordinate quantitative aspects of things already existing, but no competence to register the very reality of those things and not even the reality of their purely quantitative features.

On that resolve, which is most Duhemien, depends the proper articulation of the relation between science and faith. Duhem contributed to this project pivotal notions and precepts, but they all were meant by him to help doing physics in the best possible way. He never wanted to do philosophy as such, let alone apologetics. This, however, shows the enormous usefulness of a purely lay vocation or work, not only in the sense that a layman does it, but that it is not articulated with reference to the sacred. Underlying such a methodical separation of the two orders, lies a trust that both come from the hands of God. Only such a trust gives strength to cope with a world in which various levels and aspects of being and knowing remain conceptually irreducible to one another.

To sum up this state of affairs in the facetious phrase that what God has separated no man should join together, would be very Duhemien. He would have been the first to warn against the ever-recurring waves of reductionism, the perennial strategy of materialists of all sorts. Duhem would also be the first to warn about the dangers lurking behind aspirations that aim at an "integration" — a word never properly defined — of science and faith. All too often in recent years such integration has either petered out in well-meaning rhetoric or led to an advocacy of pantheism under various covers, such as nature's trend toward the Omega point.

That there is no trace of either in the dicta of Duhem on physics, philosophy, and faith has much to do with his refusal to abandon the realist standpoint, however unpopular. In that refusal of his — which should seem very significant in view of his friendship with such fideists as Laberthonnière and Blondel, of which more later — he must have derived conscious support from another principal article of faith, the one about the Incarnation. The proof of this is just as inconspicuous, though very specific, as was the case in connection with the relation of his philosophy of physics to the dogma of Creation.

The proof is inconspicuous as it lies covered by more than four hundred pages in the *Système du monde*. Few brave minds had the stamina to read through that volume in which Duhem dealt with the cosmological foundations of medieval Arab and Latin astronomical theories. As any other volume of the *Système du monde*, volume 4 too was full of scientific data, of texts never printed before, and of the meticulous analysis of their philosophical background. As Duhem shows in that volume, both Muslim Arabs and Latin Christians had been challenged by Neoplatonist cosmologies, all so many emanationist systems. Duhem was particularly intrigued by the failure of the Muslim philosopher-mystic, al-Ghazzali, to refuse to go along with a tenet of Neoplatonism, although it contradicted the very essence of the doctrine of creation. According to that tenet the emanation of all beings from the One, or the emanation of any lower being from a higher, takes place because of the higher being's desire for the lower. To illustrate the point al-Ghazzali referred to the desire of a shepherd for a sheep. Al-Ghazzali admitted the logical consequence of the emanationist perspective, namely, that the being desired becomes superior to the one that desires it: "The shepherd insofar as he is a shepherd [who desires the sheep] is inferior to the sheep, although superior to them insofar as he is a man."[25] Al-Ghazzali failed to note the destructive consequences of such a view for belief in a transcendental Creator who in no sense can be inferior to the beings created by Him.

If a passage is quoted, it is usually because a particular idea is expressed in it in a remarkable way. But a quotation may also serve an author as an opportunity to put forward an even more remarkable idea. Indeed, the mere sight of reference to a shepherd in al-Ghazzali's phrase touched Duhem in the very core of his outlook on existence, a deeply religious core. That it was almost a part of his nature can be seen from the naturalness with which he offered his comment that centers on the Incarnate God, the very center of Christian religion: "Assuredly, no philosophy outside the influence of Christianity could make intelligible the benevolence by which the superior being desires, without compromising his own status, the good of the inferior being. No such [non-Christian] philosophy could comprehend that . . . the good Shepherd loves his sheep to the point of giving his life for them."[26]

Here again Duhem touched upon and then left unexploited an area of enormous ideological significance. Throughout the almost 5000 pages of the 10 volumes of the *Système du monde,* he made here and there a brief, though pointed, reference to the role which Christian faith played in the rise of science. One of those remarks came as he concluded in the second volume his vast analysis of ancient Greek science. For Duhem, as well as for his friend Paul Tannery who was the foremost expert around 1900 on Greek science, the latter had long lost its vitality by the time Christianity became a noticeable social and cultural factor.[27] The historical record gave the lie to the hallowed cliché that the "otherworldliness" of Christianity made impossible the blossoming of the Greek spirit into a full fledged scientific mentality. The true reason for the stillbirth of science (an expression of this author and not of Duhem) lay in all great ancient cultures including Greece, in pagan eternalism. It had for its quintessence the doctrine about eternal recurrences or the system of Great Years. Duhem offered his comment, mincing no words:

> To the construction of that system all disciples of Hellenic philosophy — Peripatetics, Stoics, Neoplatonists — contributed; to that system Abu Masar offered the homage of the Arabs; the most illustrious rabbis, from Philo of Alexandria to Maimonides, accepted it. To condemn it and to throw it overboard as a monstrous superstition, Christianity had to come.[28]

If such was the case, one could rightly expect something new to come to science through Christianity. Indeed in 1914, when the second volume of the *Système* saw print, Duhem had already given a glimpse of the documentary proof of that novel development. He did so in in the third volume of *Etudes sur Léonard de Vinci* whose Preface (see Text 13) is a summary of the grounds which the study of medieval science provides for a Catholic to be grateful to Providence. There Duhem quoted a passage of Buridan that in itself posed a supreme threat to the prevailing secularist accounts of the origin of Newtonian science. The presentation of that text in all its details did not see light until volumes 6 - 10 of the *Système* began to be printed from 1954 on, after almost 40 years of delay and only after the threat of a lawsuit by De Broglie, Perpetual Secretary of the Académie des Sciences, put an end to an already 20-year-long footdragging by Hermann and Cie.[29] In that 6th volume one can read in full the passage which Duhem deciphered from Buridan's commentaries on Aristotle's *On the Heavens*. The passage shows Buridan to disagree with Aristotle on a pivotal point of Peripatetic cosmology in which there can be no beginning. Buridan's disagreement with Aristotle was anchored in the Christian belief in creation out of nothing and in time or "in the beginning," the conceivably most radical departure from pagan eternalism:

> Since the Bible does not state that appropriate intelligences move the celestial bodies, it could be said that it does not appear necessary to posit intelligences of this kind, because it would be answered that God when He created the world, moved each of the celestial orbs as He pleased, and in moving them He impressed in them impetuses which moved them without His having to move them any more except by the method of general influence whereby He concurs as a co-agent in all things which take place. . . . And these impetuses which He impressed in the celestial bodies were not decreased nor corrupted afterwards, because there was no inclination of the celestial bodies for other movements.[30]

In Buridan's account of the manner in which all motion had begun in the first beginning, Duhem rightly saw a clear anticipation of Newton's first law, or the law of inertial motion. Without saying much more, Duhem certainly brought into focus a most unexpected aspect of the unfathomable riches hidden in the phrase, "in the beginning," lurking behind Buridan's words, "when God created the world."

The ultimate significance of Buridan's statement was nothing less than that science could not have its beginning except as the fruit of the most fundamental of all beginnings. This certainly made sense from the viewpoint of theology, although most theologians still have to learn about Duhem's discovery, now almost a century old. From the scientific viewpoint, so often held to be antagonistic to theology, the insight of Buridan should at least conjure up a tremendous difference, the one between the circle (or wheel or swastika) and the arrow. The circle is a stamp on the scientific sterility of all great ancient cultures, all under the spell of the treadmill of eternal recurrences. The arrow, as a symbol of linearity, is expressive of a beginning which is a move in the forward direction. It is a beginning that put Western culture on the move and secured for it a pre-eminent global position.[31]

That a largely de-Christianized Western world prefers to ignore its very beginnings should not seem suprising. Nor should it cause any surprise that as its moral resources are depleting, that very world is heading towards anarchy. Duhem would find a startling justification for his historical analysis of science in that warning about an impending anarchy which came in recent years from a prominent historian of science in America. As a champion of fearless logic, Duhem would also applaud the courage with which Prof. L. Pearce Williams made it clear (in a letter to *The New York Times*) that, Donald Kagan, a colleague of his at Cornell University was completely logical for denying to civil disobedience a moral aura:

> What Kagan, I think, was arguing was that there is no 'moral' universe to which citizens can now appear that provides an adequate basis for disobedience to law. I find it strange that liberals, who insist upon the ultimate relativism of all moral values, suddenly appeal to 'higher' morality (which they are careful not to define) when it suits them. All that went out with the Victorians, and we now inhabit a society in which all moral opinions seem equally valid. . . . The point is . . . that we live in a consensual society in which we often have to do things we don't want to do, or even think are wrong, because we have agreed to abide by majority rule. Destroy that agreement, and the result is not freedom but anarchy — a condition which the United States seems rapidly approaching.[32]

It was in fact in late-Victorian times that the Western world turned its back on its Christian origins that made no sense without belief in that Origin which is the creation of all "in the beginning." But if anyone, then a historian of science should know that it was not so much the idea of evolution as the rank materialism grafted on it by Darwin that played a crucial role in that turnabout. Duhem was a firm advocate of evolution. He was enough of a realist to see the struggle of races; he wrote graphically of the United States as a homogenizing cauldron of a wide variety of peoples. But with his logical powers he could not fail to see the fallacies of Darwinism. He was deeply saddened by the fact that Darwinism as a materialist ideology became part of official university education in the Third Republic and largely under the influence of Renan.

Parading as the high-priest of Darwinism Renan sought a recompense for a disappointment in life. He felt that if he had studied biology, he would have anticipated Darwin. On a closer study of the history of the idea of evolution between Buffon and Darwin, Renan might have realized that in his *Avenir de la science* he merely repeated ideas very much in vogue for some time. But he certainly spelled out the full consequences of evolutionism taken for a Weltanschauung or religion. It implied the tenet that, as he put it, "nothing is created, nothing is added,"[33] and the fearsome consequence that, although "the universe has an obscure conscience,"[34] the sharpest human pains "mean nothing from the viewpoint of the infinite."[35] He sensed something of the irony that in an evolutionary universe even the idea of evolution loses its meaning because "philosophy is man himself, every man is born with his philosophy as with his style."[36] Renan, however, did not see the irony that if such was the case anarchy was waiting in the wing, and in fact it seriously threatened his cherished Third Republic just before the turn of the century.

Last but not least, science taken for evolutionism meant the very opposite of the unitary organic role which Renan and countless others expected it to fulfill, or to quote his very words: "To organize mankind scientifically, such is the last word of modern science, such is its audacious but legitimate pretension."[37] Such was scientism, a word still to be coined by young Maritain, in which science became the supreme self-expression of the ultimate entity, a half-conscious universe. By claiming about the science of the day that it was perfection itself, advocates of scientism deprived it from gradually perfecting itself in a genuinely scientific way.

So much in a nutshell about the unexpected portent of the unfathomable riches hidden in the phrase, "In the beginning," and about the destructive consequences that raise their ugly head whenever, and especially in a scientific age, that phrase is systematically opposed. While Duhem did not develop at length that portent, he made no secret about the epoch-making significance of Buridan's disagreement with Aristotle on the alleged eternity of the world. Moreover, Duhem noted in the naturalness of Buridan's declaration an evidence of the Sorbonne's commitment to theological orthodoxy around 1330 or so. But Duhem's principal interest lay in the bearing of Buridan's statement on the fate and future of science. As one specializing in the science of mechanics, the par excellence science of motion, Duhem knew that the birth of the science of motion was also the birth of science. All that happened in science from Buridan until today proves that science comes into its own insofar as it deals with things in motion, be they the earth or electrons, planets or photons, galaxies or gluons.

Once more Duhem did not let unsuspected opportunities for apologetics distract him from his chosen purpose, the promotion of the perfection of physics. Such a purpose, he felt, could be eminently served by unfolding the history of physics. He went about that task on a scale unparalleled before or after. With no assistants to help him, with none of the modern

research conveniences — xerox machines, microfilms, not even
a typewriter or a ball point pen — at his disposal to save most
precious time for reflection and writing, he labored in a way
inconceivable today. The least known aspect of this related to
his right hand — all too ready to tremble during his last ten
years when he filled 120 notebooks, each 200 pages, with
excerpts from almost a hundred medieval manuscripts. At first
he had to beg for them from various French libraries, especially
the ones in Paris, to say nothing of the problem of locating
them. At that time   central catalogues accessible through
computers did not yet exist even in the wildest dreams.

The aim of his superhuman historical researches had only
one purpose: to demonstrate that the truly fruitful physical
theories had always been "formalistic," that is, void of speculations about the "mechanism" of physical processes. As is well
known, he was carried away now and then through his profoundly reasoned conviction about the superiority of a formalistic method over the mechanistic viewpoint. In his little
masterpiece, *To Save the Phenomena* or *An Essay on the Idea of
Physical Theory from Plato to Galileo*, a tremendous erudition goes
hand in hand with a conspicuous failure to appreciate the "realist" motivation that drove the chief architects of heliocentrism,
Copernicus, Kepler, and Galileo. He saw too much merit in the
formalist Preface which Osiander attached to Copernicus' book.
As a consequence, and contrary to the hopes of Prof. Mansion,
editor of the *Revue des questions scientifiques*, the little masterpiece
did not dispose of the Galileo case.

Yet even there the failure was only in part. By then almost
three hundred years had gone by since Galileo's condemnation,
a sufficiently long time to see in perspective some attitudes,
including those taken by semi-official theologians of the
Church. Ecclesiastics, who kept pouring out books and articles
on the Galileo case, failed to note some all-important points.
The chief of them was the fact, still unexplained today, unless
one resorts to that explanation which is God's providential

assistance of the papacy, that in the last minute a pope, Paul V, all-too-well-known for his readiness to hand down peremptory declarations, left the decision to a mere Cardinal, Bellarmine, however eminent. Just as now, then too, no sensible Catholic, or non-Catholic for that matter, took Cardinals for infallible mouthpieces of the Church.

The other point relates to lessons that can be drawn from the merits of a particular strategy. Nothing can be so spectacularly effective as vanquishing the enemy on a battleground of the enemy's choice. In this case the ground on which many of the stalwarts of the Counter-Reformation wanted to become victorious was the literal interpretation of the Bible so dear to Luther and to his followers. The strategy imposed on its adepts a curious myopia about the great texts of Jerome and Augustine, urging caution about taking literally some biblical passages on the immobility of the earth. Galileo, who learned about those texts from his most eminent student, Benedetto Castelli, a monk from Monte Cassino, was most eager to quote those texts in his famed *Letter to Grand Duchess Christina*, that circulated widely in manuscript from 1615 on, a year before his first condemnation. He proved himself a better exegete than his ecclesiastical judges, who, however, proved themselves better judges of matters scientific than Galileo, by insisting that, contrary to his claims, he provided no *experimental* evidence on behalf of the motions of the earth. Clearly, the strategy of battling the enemy on a ground of his own choosing can at times earn but a Pyrrhic victory. This should seem to be a point with eery relevance for our times, when so many Catholics try to vanquish the subjectivism of German idealism with a "Christian" version of it.

Such a justification of long-term views would have deeply pleased a Duhem, most diffident on new fashions, intellectual or other, and always eager to find his way along perennial perspectives. Few of those scholars, whose achievements touched on more than one field, had displayed the unity of mind and purpose displayed by Duhem. Not only in physics, but also in philosophy and historiography, his labors stood in the exclusive service of one aim: the promotion of the perfection of physics as he saw it. This lay vocation germinated in his mind with no reference to any other aim and, as he had become increasingly caught up in its execution, he let nothing, not even the prospect of a most effective apologetics, interfere with his single-minded devotion to it. This is why his monumental scholarly work provides a safe foundation for a defense of Christian culture in an age of science, a defense that is far above the run-of-the-mill type apologetics or discussion of the relation between science and faith.

Duhem's great insights into the limitations of the scientific method and into the Christian matrix of the origin of the science of motion are in their monumentality forever immune to the suspicion that it is a work of "apologetics." Not that he would have been ashamed of pleading for the truth (the very meaning of apologetics) of a basic harmony between science and faith. He would have been the first to note that all intellectual endeavors represent such a plea in one form or another. But at least those who do not want to see this elementary truth will have, in facing up to Duhem's work, no justification for inventing stories such as that Duhem merely obeyed orders from the Vatican when he started his monumental researches on medieval science. (Those inventing and propagating such stories are in fact attributing to the Vatican a clairvoyance, a sort of infallibility even in historical research). But even if Duhem had been given such orders, he would have merely said that nothing is more honorable for a Catholic than to comply promptly and wholeheartedly with any such order as there may be a providential directive behind it.

As a believer Duhem was, of course, entitled to see the hand of Providence in the fact that relatively early in his investigations of the historical truth about the development of the science of mechanics, he caught a most unexpected view of the previously unknown continent of medieval science. As a believer, he had every right to thank Providence. His exile from Paris could appear to him, in the long run at least, a blessing in disguise. Had he stayed in Paris, he would have had too many students, he would have had too many contacts to handle, too many meetings to attend, too many visitors to receive — so many heavy drains on time, the most precious asset for scholarship, provided genius is on hand. Isolated as he was in many ways in Bordeaux, he had practically unlimited time when he most needed it, that is, from 1905 or so on when limitless could appear to him the area opening up before him. He must have also thought of his deep personal losses as a means chosen by Providence, so that he might devote himself single-mindedly to his providential task. That such is not a mere guess based on the wisdom of hindsight will be evident from a close look at Duhem, the deeply devout Christian.

5

The Christian

Evaluations are about values which can be true or merely perceived. In weighing the Christianity of someone, like Duhem, who lived a hundred years ago, perceptions can easily become mere projections of the present into the past. Categories, such as liberal versus conservative, that today almost impose themselves, are a case in point. These two categories, if not mere labels, did not imply for most Catholics in France in Duhem's time the meaning they convey today. To be sure, there were not a few among them who today would be classified as arch-conservatives. Had Duhem had any sympathy for them, the abbé Pautonnier would not have shared with him the revulsion he felt about Church matters as they stood in Bretagne around 1900. In the long letter which the abbé wrote to Duhem during his visit in Rennes around New Year's Day 1902, he described the Church as an institution which "is being deified here and cannot therefore reform itself. A revolution must be unleashed by God to implement that reform."[1]

By those reforms the abbé Pautonnier meant anything but what today would be called a reform of theology taken for a theological liberalism hardly different from sheer modernism. A liberalism, which today all too often means plain tampering with dogmas along a broad front, involved a hundred years ago only a handful of French Catholic intellectuals — clergy and laity. For their overwhelming majority modernism had appeared in its true nature for some time before it found full expression through the pen of Loisy. In the writings and life of Renan and of the ex-Carmelite Père Hyacinthe they had vivid illustrations of the full force of the logic of theological liberalism. Hippolyte Taine, an agnostic and a keen observer of the Church, did not expect the Church to surrender easily to modernism as he wrote in 1896: "If Catholicism resists this attack, it seems to me that it will forever   be safe from all other attacks."[2] Taine would have needed more than clairvoyance to foresee some post-Vatican II developments inside the Church.

That theological liberalism, insofar as it was modernism a hundred years ago and neo-modernism today, had no serious support in France around 1900 has many illustrations. One of them relates to Père Laberthonnière, a friend but not a particularly close friend of Duhem, who promptly submitted when his periodical, *Annales de philosophie chrétienne*, was censured by Rome. Unlike not a few   neo-modernists today, the Père Laberthonnière had no wish to practice rank disloyalty by resorting to transparent legal maneuverings whereby to postpone endlessly his submission. Failure to see this elementary point may have various sources, none of them with ties to respectable scholarship. One of those sources is a rank anti-Catholic animus which is all too obvious in some Duhem-specialists. It makes them unable to accept that unlike all forms of Protestantism, the Catholic Church even today can, in spite of enormous pressures from within and without, hold high the standards of objective religious truths and moral precepts. In the case of one of them this animus has inspired the plain absurdity of seeing in Duhem a crypto-modernist on the sole basis of "guilt by association." He had a few friends, such as Laberthonnière and Blondel, holding views which, if carried to their logical extremes, would have meant modernism. Yet even those friends of his showed an unconditional loyalty toward the teaching authority of the Church. There was nothing common between them and that true modernist, Alfred Loisy.[3]

The categories, conservative and liberal, had very limited applicability to Duhem even in that sense in which they were limited to Catholic action. What made in Duhem's time a French Catholic a conservative or a liberal depended on political attitudes. Duhem certainly was a conservative insofar as he was a Royalist. Yet his "royalism" did not extend beyond not casting his vote for Republicans, Radicals, let alone for Socialists. There is no indication whatsoever as to the political party of his preference. His sympathies for the extreme right did not go beyond reading *La parole libre*, launched in 1892 by Edouard Drumond, a virulent critic of the moral duplicity evident in many policies of the Third Republic. No more tangible were Duhem's ties with the Action Française which counted him as one of the early subscribers to *La patrie française*.

Those who with an eye on Duhem's reading those newspapers would rank him as follower of the antisemite Drumond or of the pagan Maurras would be caught in a self-defeating logic. Then they would have to take any French Catholic reader of *Le Monde* for an affiliate of the Grand Orient, to view any British Catholic reader of the *Times* as a champion of the Anglican establishment, and to list any American Catholic reader of the *New York Times* as a supporter of abortion and homosexuality. Does one become an antisemite by reading *The Washington Report on Middle East Affairs* in an America where the media had, until the onset of the Intifada, systematically kept the public in the dark about the plight of the Palestinians?

Duhem certainly wanted to be informed about facts and keep them in focus. This is why he took a "liberal" position on an issue which greatly divided French Catholics around the turn of the century. The issue was the practical policy concerning higher education. Duhem felt that whatever the need for Instituts Catholiques (which then as now were forbidden by law to use the title "University"), most French Catholics desirous to have university education had to make use of the State universities, regardless of the agnostic and at times openly anti-Catholic atmosphere dominating them. Moreover, he saw in this more than a policy demanded by practical necessity. He saw Catholic presence in State universities as a form of leaven, but a leaven to be kept true to its eminently spiritual nature.

This is why he joined forces in establishing a Catholic student organization at the University of Bordeaux only when he saw that spiritual perspective securely established. This is why he displayed a twofold attitude toward the Sillonists who became very active in Bordeaux from 1906 on and particularly in Duhem's own parish, Sainte Eulalie. As they were selling after Sunday masses copies of *Le réveil démocratique* ["Democratic Awakening"], he kept teasing them: "What a pity that you don't sell the '*Sommeil démocratique*' ['Democratic Somnolence']; I would buy a copy right away."[4] But he was ready to help young Sillonists whenever they turned to him with questions relating to the relation of faith and science.

Duhem felt, and rightly so, that as long as spiritual and theological perspectives were kept in focus, social applications not only readily followed but also retained their Christian character. He was quite distrustful about Catholic social or political action which, owing to the circumstances, had to take those perspectives lightly. Thus while Duhem welcomed Leo XIII's ruling that Catholics must not adopt the term "socialist," he viewed as impractical the pope's counsel that French Catholics become an integral part of the Third Republic. Democracy as an ideal was one thing, its actual form or practice in *that* Republic quite another.

In fact, in the closing years of the century the moral crisis of the Republic reached such heights as to justify Duhem's scathing remark: "When even the rats leave the sinking ship,

Catholics rush aboard."[5] Yet Duhem was more than ready to recognize evidences of probity regardless of their context and provenance. The best illustration of this is his letter, already utilized in ch. 2, which begins with his concise recollection of his personal experiences of the Commune. His remark about "Paris in the hands of bandits" is followed by a glowing note on the comportment of Eudes, minister of finances during the Commune, and of his wife. Eudes was "a man of probity: while he ran the ministry, his wife continued doing her laundry in the public washbasin and when he was shot down, she followed the instructions her husband had given her and took to the Versailles government the few State papers that were in his possession." Duhem rushed to draw the contrast with an eye on early 1914: "As we are today governed by bandits, the minister of finances is a multimillionaire through being a dishonest gambler and when a journalist accuses him, the minister's wife kills the journalist." Then Duhem turned to his daughter: "You see that without yet being very old I have seen things that are apt to make me nastily sceptical about politics and also as regards the definition of bandit and non-bandit."

One did not have to be a Christian to be struck by signs of decay in public life. Taine could make a strong case out of that decadence and blame for it the Revolution and its ideology.[6] Duhem was fond of repeating the point made by Taine without seeing that it could undercut a very Christian view of history. For insofar as Duhem's idealized vision of France was that of a Catholic France, it had to embody the more general truth about the Kingdom of God on earth. There the wheat always appears together with the cockle and the weeds must not, under the pretext of reform, be the target of a wholesale uprooting lest the good growth suffer in the process.

Whatever the limitations of Duhem's perspectives on Church and State, his eyes always remained wide open for opportunities where the Christian's engagement in society really proves itself. He had a deep compassion for the sick, the lonely,

and the poor or the underpriviliged as one would say today. He
translated this compassion into actions worth recalling. During
his last years his doorbell was regularly rung by a blind man led
by a young girl. Duhem was greatly surprised when one day he
saw the same man alone in the street and heard his loud
greeting. Clearly, the man faked blindness. In his first impulse,
Duhem vowed never to give him anything again. But when the
"blind" man turned up at his door next time, he relented and
gave his usual alms. "What else can I do? . . . He is a devil, but
a poor devil after all," was his reply to his daughter who
expected the "blind" man to be turned away.[7]

On many afternoons as Duhem returned home from the
University, he stopped in the nearby hospital run by the
Franciscan Sisters. There he sat by the bedside of those elderly
sick who had no relatives. The sisters found in him a guide to
the sick and the needy whom he had discovered in the
neighborhood. When he felt the call of Christian charity, he
was not sparing with money, and not even with time, that
greatest commodity of a *savant*.

Undoubtedly, a young man who was doing menial work
in the University laboratories had experienced evidences of
kindness on Duhem's part before he became the beneficiary of
that kindness that could be appreciated only from heaven.
When the young man suddenly died and no relatives showed
up, Duhem took charge of the burial and followed alone and
bareheaded the casket to the cemetery.

It was with the same immediacy that he responded to an
old man in Cabrespine, a veteran of the Italian war of 1867. He
asked Duhem whether he knew some people "up there," that
is, in Paris. At stake was the veteran's pension, the sole income
of a large family, which was suddenly cut off by those "high-
up." Duhem first thought of writing a letter to the Ministry of
War, but then he realized the ineffectiveness of the step. He
decided to write to Painlevé, although he had not been in
contact with him for some time, partly because of Painlevé's

adding political laurels to scientific ones, and partly because of Painlevé's failure to oppose Berthelot in connection with Duhem's career. To write to Painlevé could seem demeaning on Duhem's part, but write he did. The humble and friendly tone of his letter can be gathered from the fact that the pension was restored within two weeks. Clearly, it was not without reason that Duhem was called "bon Monsieur Pierre" in Cabrespine.

Three quarters of a century later, some elderly villagers of Cabrespine still spoke of him to this author during his first visit there in June 1982. They recalled not only the chocolate he used to give to the children on each Sunday as he gathered them in his home after mass, but also his teaching them catechism. He did so by filling a large-size album with drawings about the chief events of salvation history, an album that unfortunately can no longer be located. One elderly man spoke to me with vivid animation about Duhem's drawing of the Fall of Lucifer: "The angels were high up and the devils way down."

The most touching of Duhem's concern for the poor also shows his readiness to help once occasion, however unexpected, arose. In one of his hikes in the mountains around Cabrespine in the late summer of 1908 he came across a thirteen-year-old shepherdess, not only destitute but also visibly sick. Duhem could easily diagnose that Marie-Louise Gabaud, from the nearby hamlet of Trausse, showed the symptoms of tuberculosis. He persuaded the parents that their child's only chance for survival was to be taken to one of the famed sanatoriums in Le Molleau, a suburb of Arcachon on the Atlantic coast, at about 50 km southwest of Bordeaux. Once there, at Duhem's expense of course, Marie-Louise was regularly visited by him on Saturday mornings for over half a dozen years. He brought her news from her parents, talked to her about the mountains and the sheep, and saw to all her needs. Duhem was no longer alive when Marie-Louise, fully cured, entered the order of the nuns of Doctrine Chrétienne and took the name, Soeur Saint-

Pierre. Clearly, Pierre Duhem did not prove unworthy of his Christian name.

Duhem's visits to that sanatorium have a very explicit relevance to the most important implementation, which is teaching by example, of the true relation between science and faith. A good friend of Marie-Louise at Le Molleau was a young woman, a schoolteacher, who often joined Marie-Louise as she received "bon Monsieur Pierre." That Duhem was a master not only of teaching but also of preaching by example, and not by words, has a moving proof in the letter which that young woman wrote on February 25, 1916, to Duhem:

> Every day I thank God, that very God to whom you have led me back. Why should I not tell you? Much more than the exhortation of the Mother Superior in the workshop, much more than the arguments of a good priest from Bordeaux, you have made me come back to God. How? Not by your direct words, because the conversation never touched upon the subject of religion, but by your example. I have felt that you yourself accepted nothing without inquiry, that after all you reached much higher in the realm of thought than I did. And often when I am vacillating — because I shall never have the good fortune of a peaceful faith, safe from all dispute — I think of you, and say to myself: others, who know better and more than I, have believed, therefore why should I not do as they?[8]

There is no telling how many students at the University of Bordeaux must have had a similar feeling on just seeing Duhem. His mere presence must have put a damper on those ready to trumpet outworn tunes about the conflict between science and faith. The same presence must have enormously encouraged those who were still groping for a solid view about a basic harmony between religion and science. They had plenty of opportunity from early 1913 on when Duhem, after some hesitation, threw his weight behind Albert Dufourcq's project of establishing the Association des Etudiants catholiques de l'Université de Bordeaux.

Duhem stopped objecting to the project only when he saw enough assurance that the Association would not become a place for political activism under pious labels. The bylaws of the Association provided for biweekly conferences which could have for their topics subjects from social and health problems, but the weekly conference after the 10:30 mass had to be on theological matters. While he never attended those biweekly conferences, he never missed the ones on Sunday. His presence among student members of the Association received a moving portrayal by the abbé Bergereau, chaplain of the Association:

> In his extreme modesty Duhem did his best to appear a mere 'old boy'. We persisted in seeing him very young, possessed of a youth which, as we at times flattered ourselves, he recovered through his contact with us. In order to forget that his beard was white and that he even passed fifty, it was enough to see him in our midst with his bearing so simple and yet so imposing where one could notice an uncommon physical vigor, with his magnificent face which radiated the power of his intelligence, the proud independence of his character, the frank and strong goodness of his soul — with his *élan*, gayety, and untiring verve. Thus we counted on having him with us for a long time. Well, accustomed as we were to see him in our midst, not even the thought occurred to us that we might lose him one day.
>
> He was in fact present at all our meetings. Did it not occur to him to leave Paris on Saturday evening, depriving himself of one more day which he could have spent with his daughter, in order not to miss our mass the next day? In our study group, where especially during these recent times he made it his duty to assist, with what exquisite benevolence did he listen to the poor lecturers as we are, and to each question how many appropriate, luminous, and delightful things he could always add by tapping the

treasures of his knowledge! Who has not admired him, as the days of communion came, kneeling down side by side with us at the Holy Table? At all our banquets he had his place and how quickly he became there the Normalien, and that impish mind which he once was! You have not forgotten the delightful companion he was at Verdelais on the day of Ascension. With a joyful and brisk step, laughing with one, arguing with another, on the road from Langon to Verdelais, he climbed the banks of the Garonne in the morning and then descended in the evening. Lucky the one who could maneuver himself to his side! Everybody wanted him to himself alone. I had to admire on seeing with what simplicity this great man, by eliminating without any fuss the distance between himself and others, became all to all and succeeded in being for you another *camarade*! He succeeded so well that at times you did treat him as one.

I remember that several times some of you, after having entered the meeting room before Sunday mass and surprised at not having noted him immediately, simply asked: "Where is Duhem?" Certainly, Duhem was not far. He was indeed so close that he happened to hear the question. Red color rushed into the faces of these daring fellows. But Duhem smiled contentedly. One sensed that he was happy. He knew well that these little familiarities took nothing away from the respect and high regard we had for him in our hearts. It is not astonishing that he became the soul of our group to the extent that, in speaking of us, the public began to say: The work of Monsieur Duhem.[9]

On all those weekly gatherings Duhem let the students freely offer their views and questions. His few formal addresses were a marvelous combination of wit and depth, intellectual as well as spiritual. They reflected his conviction that the ultimate factor which makes one see harmony or conflict between science and religion goes beyond purely intellectual matters. An exquisite illustration of this is the address he delivered on June 14, 1914, at the annual banquet of the Association. It followed a toast by Prof. Arnozan, of the Faculty of Medicine, who urged the students to be intent on their career. It now became Duhem's task to explain the all-important difference between what Prof. Arnozan meant and careerism. In the latter Duhem saw a chief plague of academic and professional life and a mainspring of fallacious reasonings whereby a Christian intellectual justifies his siding with those who claim that faith and science are irreconcilable. Such cases were no less plentiful in Duhem's times than they are today. After having spent by 1914 a quarter of a century in the secular academia, Duhem had inside information that must have been eye-opening to his youthful audience. Hardly any of them was unaware of the many painful aspects of Duhem's career and the moral uprightness with which he coped with his academic trials. As a result, his speech did not sound like specious moralizing but a message that carried an authority which only a much tested life could provide:

> One evening about thirty years ago the pillars of free thinking were assembled in Taine's home. Marcelin Berthelot chatted with the host. Leaning on a table, the corpulent Renan scribbled something on the end of a sheet of paper. The sketch showed a tombstone with the epitaph: Here lies Berthelot. He occupies the only place which he never requested.
>
> My dear friends — Duhem continued — don't desire each and every place. When a place becomes vacant, you ask your conscience: To fill that place am I the man who is needed, the right man in the right place? And if your conscience tells you no, you should not go forward.
>
> If your conscience tells you 'yes', you cast a glance around. You will search whether among your fellow applicants there is not one who is worthier of the post which is

your ambition. If you notice one, you will yield so that he
may advance; indeed, I say, you will help him advance.

If you have recognized deep in your heart and conscience that you are the most worthy of the post, you will
still forbid yourself any means which cannot be used in
broad daylight, any procedure which would not be of the
most scrupulous loyalty.[10]

Duhem readily acknowledged that his counsels had a
practical sense only inasmuch as anyone obeying them would
be saved of having to listen eventually to the reproachful voice
of conscience:

Undoubtedly, I am overwhelmed with commissions
and loaded with honors: but while people smile in my face,
I hear them laugh behind me. I know they call me incapable, a nullity. . . . True, I am worthy of the rank I occupy,
but such and such was worthier than I, and because I have
vanquished him he is poor, humiliated, unfortunate. . . .
You were worthy of the place you occupy, you were the
worthiest. But in order to obtain it, what platitudes, what
dirty tricks, what apostasies were necessary.[11]

Such was a perspective in which alone, Duhem felt, the
apparent contradiction between Prof. Arnozan's advice and his
own warning against careerism could be resolved. Within that
perspective even a toast could be turned into a meditation, in
fact, into a prayer:

Do you believe, my dear friends, that the happiness of
escaping all such reproaches, the pride of holding high
one's head without blushing, would not be sufficient
consolation for some disfavors and injustices? Do you feel
that you will receive your recompense in this world? What
should I say of the recompense which is waiting for you in
the world to come? Well, being careerists, though better
advised than Berthelot, you will have requested the only

place which is worth the effort, because no reversal will deprive you of it as one holds it for eternity. . . . When one is a Christian, when one does not believe in luck, when one believes in an appeal to Providence, a good wish is a prayer. Do not be astonished that my toast takes the form of a prayer. As you take your place at the banquet of life, I implore God to go along the tables often enough to notice the humble place where your modesty made you sit down and, taking you by the hand, tell you: My friend, move higher up.[12]

Two years later, when Catholic women students celebrated the first anniversay of their own association, Duhem was asked to deliver the festive address. There he gave a glimpse of the most elevated perspectives in which he saw the nature of women, their professional training, their function as wives and mothers united in the love of country and in the service of God.[13]

By then, or June 1916, France was going through Verdun, her most memorable trial of the war. Duhem, who had already rushed to the defense of the French mind by giving a series of lectures on German science,[14] generously lent his support to war orphans and war widows whose number grew disturbingly high. Help was needed to administer help with utmost tactfulness. Bordeaux turned to Duhem to be the spokesman of the regional committee set up to aid the victims of war. The problem he had to handle consisted in implementing an imponderable aspect of the "union sacrée." He had to assure those (and they were far more numerous than Clemenceau would have admitted) for whom the sacred meant the supernatural, the genuinely Christian. He had to allay the fear of mothers, not yet influenced by fashionable unbelief, that their half-orphaned children would not be exposed, under the guise of aid, to convictions at variance with their parents' religion. The task was all the more delicate because the "sacred," as understood by secularists, was to be spared of well-deserved strictures. As an observer later recalled, "Duhem knew with infinite tact and with very fortunate words, how to convince them that they could without worrying accept the material and moral aid offered to them."[15]

This was in July 1915. In the Fall of the same year, as aid was distributed to war orphans and new ones were inscribed in the program, Duhem was there as frequently as possible. In the same observer's words:

> Each time it was a new case of astonishment and admira- tion to see him receive with the refined courtesy of a man of high standing the widows coming to enroll in the Committee's program, putting in writing at their often unskilled dictation the information requested, and sending them away with such words of encouragement and comfort that none of the widows inscribed through his efforts had ever forgotten that 'monsieur si bon' who had received them.[16]

The Christian modesty and tactful charity of the great savant were no less visible in Duhem's contacts with the Atelier Sainte-Agnès in Thiais, in the suburbs of Paris. His daughter, Hélène, settled there permanently from 1910 on as a paying guest to help Mlle de la Girennerie, the head of the Atelier, in providing home and training for young working women. Part of the Atelier's revenue came from benefit sale of clothes made on the premises. Duhem's contributions were numerous and always offered in a way in which he made himself appear almost the beneficiary.

The poor whom Hélène befriended in Thiais could count on her father's friendship. On being told by Hélène that Gino, the little girl of a poor couple in Thiais, needed good country air in Cabrespine, Duhem let her spend a summer there. Back in Thiais, little Gino could not talk enough of "Monsieur Tic-Tac," that is, Duhem, who thrilled her by the sound of his big pocket watch. In writing to Hélène, Duhem added a postscript:

My dear little Gino. It is very kind of you to think of your
old friend, Tic-Tac. He often thinks of your big dark eyes.
He thinks that you are trotting around in the house and
bring there the happy sunshine of your laughter while it
rains outside. He would like to be near you and take you
in his big arms. Give him your little hand. He will breathe
a kiss on it and say: Gino, eat well and sleep well in order
to have big rosy cheeks. Tic-Tac.[17]

Such acts of kindness came from one who lived a truly
spiritual life. Only that life reveals itself in words that, however
plain, strike by their spiritual depth. An example is the New
Year's greeting he sent to his daughter on December 30, 1910:
"These wishes I deposit at the feet of God so that He may
transform them into realities. I do not enter into details. I sum
them up in this word: Your happiness. May God grant you all
that He, in His wisdom, judges good for you. Me, a poor man
with short views, I am not in the position to know whether
what I think to be useful for you might not indeed be your
misfortune."[18]

Clearly, Duhem's belief in Providence was the kind of
belief that results in a genuine abandon to God's will. No
wonder that he quickly discovered Thérèse of Lisieux. Duhem,
always eager to tell his mother and daughter about his discoveries, now appeared to be even more eager to communicate his
findings. What he wrote to his daughter on June 2, 1910, about
Thérèse's L'histoire d'une âme, will be best appreciated if seen
against his report to her concerning his having read "from
cover to cover" Fouard's Vie de Saint Pierre. The latter he found
"extremely interesting," an "erudite work, not at all German."
Of a different kind was his reaction to the book whose author
was to be canonized within fifteen years:

Yesterday I finished [the book of] Soeur Thérèse de
l'Enfant-Jésus. I am under the spell of that small book. It is
marvelous to see how she shows you a way along which,

without becoming a Carmelite,   without leaving the profession where God has placed one, one can become a great saint. There are in that book thoughts of admirable depth on the practice of charity. We must take that little book to Cabrespine and read it again under the chestnut trees of Granel.[19]

Earlier references in this book to the lay vocation which Duhem saw in his pursuit of the perfection of physics may now appear in their strongest light which is the light of Duhem's very Christian soul. In the spectrum of that light, a spiritual light that is, a chief trait was the simplicity and the childlike character of his faith. One aspect of that character, a most Gospel-like trait, was his fondness for traditional, well proven forms of piety. At Christmas night in 1910, after he had spent almost three hours in church, attending two masses, he was unable to fall asleep, half-frozen as he was. He decided to write to his daughter about the liturgy in the nearby church of the Franciscan nuns: "They parted with their mass where Kyrie, gloria, credo were a potpourri of popular, nay vulgar tunes; instead, they chanted a mass to tunes just as bad but less ridiculous. During the second mass we had [the old tune] of 'sweet star, oboe, bagpipe' and other old memories." His letter was finished in the evening after he returned from solemn vespers in the Cathedral.

To a man of so devout a life it was a pain to find well-meaning Catholics show an "enlightened" attitude toward miracles, apparitions, places of pilgrimage etc. The contents of the letters he wrote to his daughter on June 2 and 3, 1912, should silence once and for all those who try to portray him as a crypto-modernist. There he spoke of the discomfort he felt when in a friendly company he heard two very good Catholics, both historians, one of them a priest, picking apart a great many pious legends. "The House of Loretto, Rocamadour, scapular, the saints of Provence — all were gone over. Segan [another guest and also a historian] and I did not say anything. Segan because it is his custom, I because I do not share their manner of seeing things and also because I do not have the necessary information to dispute them." Then he added: "You would not have been happy." In the earlier letter he described the style of the ecclesiastic as one that befits "a modernist and a republican."

Obviously, Duhem did not long for a fashionable and easy Christian faith and life. His life had too many trials to let him entertain illusions, especially their spiritual kinds. At the center of his religious life stood the cross of Christ. A proof of this is his obvious identification with two crosses in the outskirts of Cabrespine, the subject of two exquisite drawings of his. Ultimately, they are the most genuine context for putting Duhem on the scene of his life and work.

One of the crosses, the Croix d'Estresse (the cross of distress), he drew on September 4, 1912. His drawing of it has its own value for students of the history of art, as the cross is a rare example of crosses with a Pietà carved on their reverse side. The cross, erected in 1632, has since attracted many pilgrims.[20] They still keep going to the place where it stood until about six years ago when it mysteriously disappeared while a new road was constructed to the property acquired by some from abroad. (Perhaps through this reference the Department of Aude will take note and appropriate action). Let it be hoped that Duhem's drawing of that cross will not become its sole detailed evidence and a painful reminder of widespread illegal trafficking in art objects in the region. In any case, the drawing by Duhem remains a lasting evidence of his spontaneous recourse to the Virgin invoked as the mother of all afflicted. It should not be difficult to evoke Duhem's sentiments as he drew the figure which in a kneeling position under the Pietà raises his hands in supplicant prayer towards the One of whom it was never heard that anyone turning to Her would have had his prayers unanswered.

The other cross, erected in 1638, a plain one in the midst of the communal field, Duhem drew on August 21, 1916, less than a month before his death. He made that simple cross speak by emphasizing its size. He did so by letting it be seen from an angle whereby it appears equal in height to the mountain behind and thus dominates the field. A purely artistic technique, but hardly in the case of Duhem who never pretended to show what he was not convinced about. He let his whole life be dominated by the cross, the very act that alone makes a Christian for whom "every treasure of wisdom and knowledge is deposited in Christ" (Col. 2:3). It was through identification with Christ that Duhem's vast knowledge of science, including its philosophical and historical dimensions, took on a prophetic character.

6

Prophet without Honor?

Duhem's identification with Christ the crucified implied a variety of crosses he had to bear: a chronic stomach ailment contracted in early life, the death of a sister, of a brother, and above all of a dearly beloved wife and of an infant son, the thwarting of his career by officialdom, various reprisals for professional excellence, and a solitary life during his last ten years. Among his consolations he could count the fact that some Catholic intellectuals showed enthusiastic appreciation of his work. They were mostly connected with the Société Scientifique de Bruxelles and the editorial board of the *Revue des questions scientifiques*. Its director, the Père Thirien, made Duhem feel assured from their first encounter on that his writings would be published with all possible speed in the *Revue*. The encouragement that can be derived from such an assurance will fail to be appreciated only by those who never had their manuscripts rejected on patently unjustified grounds.

Enthusiastic appreciation by some Catholics also played an important part in the eventual appearance of the *Théorie physique*, published first as a series of articles in the *Revue de philosophie* that had a priest, E. Peillaube, for its editor. Indirectly, the appearance of the *Théorie physique* owed much to the quick publication in the *Revue des questions scientifiques* of essays

Duhem wrote mostly while in Lille, in which he first tried out several of his basic philosophical ideas. It was again the prospect of quick publication that spurred Duhem to writing, in about eight subsequent installments in the same *Revue*, another masterpiece of his, *Les origines de la statique*, the most revolutionary or rather epoch-making work in the historiography of science. It is still to two other Catholic scholars, G. Radet and E. Bouvy, of the University of Bordeaux and editors of its *Bulletin italien* and *Bulletin hispanique*, that the learned world, but especially the Catholic world of learning, owes another epoch-making work of Duhem, the three volumes of his *Etudes sur Léonard de Vinci*. It was also in those two *Bulletins* that the material of several of the volumes of the *Système du monde* first saw print.

Catholic appreciation of Duhem's work was centered on some perceptive leaders in the Société Scientifique de Bruxelles. They were the ones who made him an honorary member of the Société in 1900 and asked him to deliver the opening address of a three-day celebration of its silver jubilee the following year. It was through the Société Scientifique that some Catholic intellectuals abroad took notice of Duhem. The result was a major article on the history of physics by Duhem, published in 1911 in the 14-volume *Catholic Encyclopedia* in New York. It also contained four short articles by Duhem, two of them on Jordanus de Nemore and Oresme, both pivotal figures in the new perspectives established by Duhem on the historical origins of Newtonian science. Neither of those articles created an echo among Catholic intellectuals in the United States where a huge Catholic system of higher education was beginning to be established.

No small blame for this lies with the continental, and especially with the French Catholic intellectual scene which was closely watched from America for important new developments. No article on Duhem has, for instance, ever appeared in the *Etudes*, the leading French Catholic monthly. French Catholi-

cism bears particular responsibility for ignoring Duhem's passionate appeal, made in 1911 to the Père Bulliot, dean of the
Faculty of Philosophy at the Institut Catholique in Paris.
Duhem's appeal had for its aim the establishment of two new
chairs at the Institut, one for the philosophy of science and
another for the history of science.[1]

Those familiar with and concerned about the extent to
which agnostics, Freemasons, and atheists have during the past
30 years expropriated to themselves such chairs, together with
the editorial offices dealing with scholarly publications relating
to those chairs, should find an eery timeliness in the words that
begin Duhem's letter to the Pére Bulliot: "As one living among
those who profess doctrines contrary to ours and therefore well
placed to know their plan of attack against us," Duhem called
his friend's attention to the fact that those in question were
especially eager to exploit the philosophy and history of science
for their purposes. Their ultimate aim is to secure a "scholarly"
platform from which "to deny, in the name of science as such,
to all religion the very right to exist."

Their principal message in teaching those two subjects —
and many present-day philosophers and historians of science
duly echo that message — aimed at making any and all
subscribe to "an agnosticism for which all religion is a dream,
more or less poetical and comforting." Their central strategy
was to gloat over the alleged darkness of the Middle Ages and
then "to make shine into our eyes the glories of the Renaissance where minds, liberated at long last of the yoke of the
Church, have found again the thread of scientific tradition.
. . . They delight in contrasting from the 16th century on the
always ascending march of science with the ever deeper
decadence of religion; . . . this is what is being taught in a
number of chairs, this is what is being written in a multitude of
books."

Duhem wanted the teaching to be done from those two
chairs to be such as to give the lie to that agnostic message. In

fact, he wanted nothing less than to have in those chairs Catholic scholars courageous enough to stand up and decry that message as a plain lie! In his time courage among Catholic intellectuals was not as rare as it is today. Partly because of this there was a great interest among them to do apologetics. If there was any problem in that it was only because, in respect to questions relating to science, and especially to questions dealing with the relation of science and faith, zeal all too often outpaced competence and proper training.

Duhem saw this on a vast scale and at close range when in early September 1894 he participated at the Third International Congress of Catholics in Bruxelles. The over two thousand participants were exposed to a number of papers on science and faith, papers that gave better witness to good intentions than to sound competence. At the promptings of some present Duhem asked for permission to make a few remarks. Although they were not reported in the form of direct quotations in the *Procès verbaux* of the Congress, the clarity and incisiveness of the text suggest that Duhem's remarks were faithfully rendered. Duhem, the report begins, "is convinced that these researches [which have for their object the interface of the positive science and metaphysics] will, if done wisely and prudently, lead to the reconciliation of Christian philosophy and modern science, but he insists on the extreme difficulty of such studies."

As his reasons Duhem first pointed out that one could not gain a proper grasp of a science by reading a few popularizations of it, be they from the pen of first-rate scientists. The mastery of science, as of anything else, demanded its assiduous practice:

> One does not comprehend the meaning and bearing of the principles on which a science rests except when one has studied that science for years, applied in a thousand ways those principles to particular cases, and mastered in depth the technique of what the Germans call the materials of science.[2]

Then, after illustrating his point with a reference to the proper grasp of Euclid's parallel postulate, which demanded "years of work," Duhem urged that those aiming at a synthesis of religion and science should "begin with studying the latter for ten, for fifteen years" and for its own sake and "without seeking to put it in harmony with such and such philosophical assertion." Duhem might have erred in thinking that such a scientific preparation would readily yield on reflection philosophical conclusions that "accord with true philosophy." He might have also overstated the length of time to be devoted to scientific training by the one determined to handle competently the interrelation of science and faith.

He may not have struck a proper balance as he contrasted the exactness of scientific answers with the comparative vagueness of philosophical propositions. Since reality as such and not its quantitative aspects was the object of philosophical reasoning, philosophers were not necessarily to imitate scientists in the manner Duhem suggested: "Once a problem is posed, scientists devote centuries, if necessary, to solving it. They accept only a precise and rigorous solution." To follow such advice in philosophy would have contradicted Duhem's own insistence on the primacy of good sense. Duhem was very much on target with respect to a question of strategy which he illustrated with a portrayal of the opposite camp. It was full of first-rate practicing scientists who took full advantage of their professional excellence to spread wrong philosophies of science:

> The schools we are combatting give us example. The positivist school, the critical school, publish numerous works on the philosophy of science. These works carry the names of the greatest names of European science. We cannot triumph over these schools except by opposing them with researches done by people who also are masters of the positive sciences.[3]

Among those names Duhem certainly included that of Ostwald and of Mach whose success in marketing their anti-metaphysical message largely rested on their being prominent scientists. But what could Duhem mean by triumph? Did he really look forward to a sweep of faith over disbelief? Would this not have revealed some naiveté on his part especially in view of his being so steeped in history? Was not the ardent hope of the coming of God's Kingdom to be tempered by the acceptance of the continual growth of weeds even within that very same Kingdom? Should not the triumph rather be specified as the strengthening of the self-confidence of the Catholic camp? And if such were the case, would it not have been necessary to call attention to the proper philosophical and theological training (again a matter of serious studies for over many years) on the part of the Catholic scientist if he were to speak in a reliable manner of the harmony of his very science with faith?

One could only wish that Duhem had faced up to these questions. But here too, as in other respects of the bearing of his researches and reflections on the harmony of faith and science, he restricted himself to some strongly worded points. And he was undoubtedly pleased for having done so as shown by the letter he sent from the Congress to his mother:

> Yesterday, I decided to create a big stir. It was in the section of philosophy. The room was full, mostly of ecclesiastics. One brave ecclesiastic had just discussed an objection taken from mechanics. My opinion was asked concerning the scientific part of the problem. Then, I bluntly told all these good Catholic philosophers that if they obstinately continued talking of science without knowing of it a single word, the freethinkers would hold them up for ridicule; that in order to speak of questions where science and Catholic philosophy touch one another, one must have done ten or fifteen years of study of the pure sciences, and that, if they had not become men with deep scientific knowledge, they must remain silent. . . . The

idea, once launched, will advance; all afternoon one spoke but of this at the Congress. I do not regret having come. I believe that the seed which I sowed will germinate. It is for the first time that those brave people heard the truth spelled out. This does not surprise me much, but I am surprised to see that they respond, or at least several of them do so with a great deal of good will.[4]

Duhem would today address no less pointed remarks to Catholic theologians and philosophers who steep their discourse on faith and science in speculations bordering on pseudo-mysticism. He would certainly expose their reluctance to do apologetics as a failure of nerve, as a failure to recognize that any respectable intellectual stance is an apologetics, that is, a plea on behalf of conclusions taken for truth and not merely for subjective opinions.

To those who would resent his strictures of incompetence, however well-meaning, as indicative of the expert's arrogance, he would rightly recall his readiness to help theologians most intent to learn about the scientific side of their topics. He received with open arms in his home in Bordeaux the newly ordained Dominican, Reginald Garrigou-Lagrange, who might have noticed Duhem at that Congress in Bruxelles a year earlier. Duhem recognized in the young Dominican, who started out as a medical student at the University of Bordeaux, a most serious readiness to go to the bottom of fundamental questions, such as the nature of motion.

That meeting of Duhem and Garrigou-Lagrange was followed by a correspondence over almost two decades which had for its principal subject the difference in which inertial motion is seen by a physicist and by a Catholic philosopher-theologian.[5] The physicist's account of such a motion conflicts with the philosophical insight for which motion is expressive of real change and therefore a surplus in every instance that must have a cause. For as Garrigou-Lagrange rightly saw, the Thomistic proof of the existence of God based on the reality of motion would be undermined if an un-ontological interpretation, the one demanded by the methods of physics, exhausts all the meaning of inertial motion.

Duhem's final clarification, a long letter written in 1912, on the difference in which the same physical reality is seen by a physicist and by a philosopher pleased enormously the by then prominent Dominican, teaching in the newly established Angelicum in Rome. First, he extended a standing invitation to Duhem to the Angelicum, second, he asked and obtained Duhem's permission to reprint that letter as an Appendix in the work, *Dieu: son existence et sa nature*, which immediately upon its publication was hailed as a classic. Through the fourteen re-editions of that work that letter of Duhem has indeed become his most often printed writing.[6]

Although read by countless Catholic theologians and philosophers, that letter of Duhem did not create enough interest in most of them to explore his vast work, a storehouse of information and insights to anyone seriously interested in the relation of faith and sciences. The opportunities lost thereby have some telling illustrations. One may take, for instance, the work, *Science et apologétique*, the text of six lectures given by A. de Lapparent, a geologist member of the Académie des Sciences, given at the Institut Catholique de Paris in May and June 1905. Did its contents really do as much good as suggested by its twelve editions in nine years?[7] Lapparent made much of Duhem's insistence on the revisability of physical theories as set forth in that series of articles in the *Revue de philosophie* which formed the first part of the *Théorie physique* published a year later, in 1906.

Lapparent failed to report the positive part of Duhem's ideas on physical theories, namely, that they gradually approach a true mirroring of the workings of the physical world. Consequently, Lapparent's apologetics became tantamount to a new version of the long-discredited method of invoking God to fill the gaps in scientific knowledge. And since those gaps keep being filled at an accelerated rate, the argument can work as a boomerang by making it appear that ultimately God is a superfluous hypothesis. Curiously, it escaped a Frenchman that it was a French scientist, Laplace, who, as a physicist, had some justification for his memorable remark: "Je n'ai pas besoin de cette hypothèse."

By contrast, it is rather ironical that in the 1930s the work, *L'Eglise et la science*, by Louis de Launay, also a geologist member of the Académie des Sciences, saw only one edition,[8] although far superior to the book just discussed. To complete the irony, de Launay's book is known to exist only in two copies in United States libraries. A mere look at its table of contents suggests a broad and balanced use of Duhem's findings, conceptual and historical, that have a bearing on the relation of science to theology. At the start de Launay discussed the difference between what faith (Church) and science stood for. Following that methodological clarification, de Launay devoted the first part of his book to a historical survey of the relation between science and faith. There the very titles of the opening two chapters — "Antiquity and the Middle Ages" and "The Renaissance, Leonardo da Vinci, Galileo" — suggested a reliance on Duhem's findings. Duhem's ideas were also in view through the very titles of two chapters of the Second Part: "The two modern mentalities" and "The measure of usefulness, limits, and uncertainties of science."

De Launay's book did not represent the last noteworthy utilization in France of Duhem's work for a discussion of the relation of science and faith. The one which Albert Dufourcq worked out somewhat earlier did not prove effective, buried as it was in his multivolume interpretation of Church history, *L'avenir du christianisme*, a work still not fully appreciated.[9] A sad contrast to the efforts of Dufourcq is offered by the thick volume, *Apologétique*, published in 1947. In its section on science Duhem was completely ignored. The sole mention of Duhem in that book is in its section on cultural history that dealt with the emancipation of Europe from the tutelage of the Church during the 14th century: "Thus, for instance, one shall see the sciences trying to invade the domain of faith. Yet, one cannot recall often enough, as this was done by Duhem, that the very structure of the exact sciences could not have developed outside Christianity. It is sufficiently evident that the sciences have not arisen outside Christian Europe."[10]

The statement was too generic to spark curiosity. At any rate, very few of the many who had that thick book in their hands could easily come across that statement in a book of 1386 pages. Certainly not one among Catholic intellectuals in the United States, where that book was available in many places, was made to feel the need for a serious reassessment, say, of history teaching in Catholic places of higher learning. Yet the statement should have appeared as startlingly novel in comparison with the book, *Catholicism and the Progress of Science*, brought out in 1940 by no less prominent a publishing house than Macmillan in New York.[11] Its author, W. McAgar, listed in the preface three well-known priests — a Jesuit, a Dominican, and a Benedictine — as ones who had read the manuscript and made valuable suggestions.

All three were, in all evidence, completely unaware of Duhem. Otherwise, the book's first chapter on the historical development of science would not have been a full three decades, if not more, behind scholarly times. The author failed to mention Duhem, let alone his heroes, Buridan and Oresme, although he often quoted Thorndike's eight-volume work on the history of magic and science, a work not silent on Duhem.[12] To crown the comedy the author quoted Whitehead's account of the origins of modern science as a subconcious derivation from medieval thought.   Beneath that comedy there lurked an intellectual tragedy. Why is it, one may ask, that English-speaking Catholic intellectuals failed to be stirred by Christopher Dawson's short but pointed references to the importance of Duhem's findings about medieval science? Why is it that Dawson himself failed to see that Whitehead's crediting of the medievals with the origin of science was a damning with faint praise? The first chapter, "The Origin of Science," in Whitehead's famed *Science and the Modern World*, which following its original publication in 1926 sold in more than a million copies in ten or so years, should have been the target of Catholic scrutiny from the very moment of its appearance in print. This should have become the case even if that chapter had contained but the phrase "faith in the possibility of science generated antecedently to the development of modern scientific theory, is an unconscious derivative from medieval theology."[13]

That phrase should have appeared for what it truly was, a "damning with faint praise," to anyone pondering, however briefly, the phrases that preceded it: "I am not arguing that the European trust in the scrutability of nature was logically justified even by its own theology. My only point is to understand how it arose."[14] Clearly, Whitehead had an axe to grind against medieval Christianity as well as orthodox Christian faith. Since that faith and that Christianity were rooted in belief in a personal Creator who did not have to create a universe, Whitehead, a champion of a pantheism in which God himself is evolving, had no choice but to slight that faith. Did Catholic readers in English-speaking countries, or in France for that matter, notice that in that book Whitehead, although often speaking of God, never referred to Him as a Creator? Did they ever ask on what grounds did Whitehead treat a startled audience to the idea of a medieval origin of modern science?

Since the idea was not in the air in 1925, and certainly not at Harvard University, where Whitehead delivered his book as a series of lectures, he must have relied on some recent studies which at that time could only be the publications of Duhem. Having for wife a French woman, who retained at least a fond memory of her convent education, Whitehead had one more reason to display keenly the professed interest of British academics in French scholarship. Following his death in 1916 Duhem was prominently mentioned in leading British newspapers. For that reason alone a more than cursory knowledge of Duhem's work may be assumed on Whitehead's part. In that case there arises the legitimate question whether Whitehead wanted to draw attention away from Duhem by offering a glittering sop, readily swallowed by Catholics, but still a mere sop, so as not to offend the agnostics, about the medieval origins of science.

Whatever answer to this question may be forthcoming from a careful study of Whitehead's manuscripts and correspondence, there can be no doubt about one point. Within a generation or so Catholics offered their full assistance to the "scholarly" procedure of damning Duhem with faint praise. As they became increasingly uncertain about themselves from the mid-century on, Catholic intellectuals readily jumped on the Koyré-bandwagon as it started rolling, following Koyré's first lecture tour in American universities. Suddenly his *Etudes galiléennes*, first published in 1940, became the Bible among younger American historians of science, soon to be the leaders in their field. Koyré's *Etudes* appeared to them as opening up exciting vistas and possibilities.

One of those possibilities derived from Koyré's casting the history of science into the moulds of mental mutations, as if the mind consisted of genes.[15] Darwinism could now appear to reign supreme, for if Koyré was right, it was applicable to that citadel of pure reason for which science has been taken in some quarters. Another "exciting" possibility lay in Koyré's claim that whatever the medieval origins of modern science, it had to be followed by a mental mutation during the Renaissance so that Galileo's and Newton's science might arise. Had Koyré's *Etudes* contained a name-index, it would have revealed the fact that he mentioned Duhem over fifty times, far more often than anyone else. This might have alerted even the unwary about Koyré's true aim. It was natural for one, who exchanged his Jewish religious heritage for the pantheism of Boehme and Spinoza[16] and for the rationalism of the French Enlightenment, to realize the magnitude of the threat posed to secularism by Duhem's researches and to oppose it by every means of sophistication.

It was also natural that French Catholic intellectuals active as historians and philosophers of science should yield to the lure of Koyré's message. Of course, they did so in the name of scholarship while they also saw the new intellectual fashion offer more than one opportunity for promoting one's carrer. The section on medieval science, which G. Beaujouan contributed to the four-volume *History of Science*, edited by R. Taton, did rank injustice to Duhem's achievements.[17] Koyré, who contributed the next section on Renaissance science, was thereby saved, and by a Catholic and a priest, from appearing too hard on Duhem. Koyré did not have to worry about being taken to task by Catholic scholars in the field.

When after almost forty years of delay, the second half (five volumes) of Duhem's *Système du monde* began to be published, no Catholic cared to expose the conspiracy lying behind the delay, although at that time personal depositions could have been taken from most authoritative sources. They included the Perpetual Secretary of the Académie des Sciences, and the director of the Centre Nationale de Recherche Scientifique. It was an open secret for at least those who wanted to take note that only the Secretary's threat, delivered in writing, of a formal lawsuit against Hermann and Cie stopped its by then a more than twenty-year-long footdragging.[18]

Yet even the full publication in 1959 of the *Système du monde*, undoubtedly the greatest individual work of original scholarship in modern times, failed to stir Catholics in France. The hundredth anniversary of Duhem's death in 1961 passed with hardly a notice. The only notable recall of Duhem in 1966, the fiftieth anniversary of his death, was an article published in *Physics Today* which also appeared in French translation in the *Revue des questions scientifiques*.[19] It was written by an American thermodynamicist who, creditably enough, made no pretense to expertise in philosophy or in history. As a Jew with markedly liberal views, he cannot be blamed for summing up Duhem the man as a Royalist with extreme religious views.

Non-Catholics, and inded non-Christians and plain agnostics, if not worse, make up the great majority of those writing about Duhem — and not always equipped with the proper scholarship required for the task.[20] Their interest in Duhem is more than a matter of pure scholarship. They seem to be intent on undermining the witness which Duhem, by his personal probity and intellectual greatness, brings againt secularism. They seem to perceive the paramount importance of that witness in an age where science and its interpretation, philosophical and historical, bear heavily on almost any issue. Their purpose can effectively be served by spreading a distorted image about Duhem as if he had been a sort of a schizophrenic: extremist Catholic conservative on the one hand and a radical positivist on the other. Emphasis on the former trait would undercut his principal findings about the medieval origins of modern science. The other trait would make him an unwitting ally of such militant anti-Catholics as Ernst Mach and several principal members of the Vienna Circle, together with their disciples among the paradigmists and the Popperians. They all are one in claiming that metaphysics is nonsensical and that intellectual history owes little if anything to Christianity in general and Catholicism in particular.

The failure of Catholics to see the crucial importance of Duhem in this great ideological battle makes for a telling, and a rather scandalous story,[21] too long to be outlined here, however briefly. Part of that story is the indifference, if not plain miscomprehension, with which they handle such a paramount witness on behalf of Duhem as the considered judgment of Annaliese Maier, the foremost historian of medieval science around the middle of this century. She concluded that "Duhem was essentially right,"[22] whatever the particular corrections to be made to his massive investigations.

Catholics are still to resonate to that most momentous judgment. But it seems that in this connection too Catholics are their own worst enemies. They had better come to their senses. Of course, in this age of anti-triumphalism among Catholic intellectuals — clergy and laity — there is no point in recalling Duhem's hope for a triumph through a scholarly exploitation of the true history of science. But perhaps in these times, when the Church experiences growth only in the Third World while she suffers huge self-inflicted setbacks in the First or Affluent World, the evocation of the specter of future disasters may serve its purpose. The specter in question may be conjured up by a plain look at what is going on in the opposite camp.

There the study of the history of science, cast in the moulds of this or that version of Darwinism, is used as a supreme evidence of the claim that ultimately all is self-evolving matter and this is all that should matter. Supporters of that camp are those historians of science who plainly turn their subject into a materialistic propaganda. They are aided by those for whom the study of the history of science is a mere technique or method, preferably free of any ideology. And if ideologies cannot be avoided, they are reduced to social consciousness or cultural patterns and the like, all void of any objective truth. That Protestant historians of science find it difficult to appreciate the medieval origins of science is understandable. The Reformers and their followers had a vested interest in portraying the Middle Ages as dark as possible. But no less understandable should seem the unwitting support which they give thereby to that camp most hostile to them as well.

Further support to that camp comes from those who, whatever their better judgment, do not dare to dissociate themselves from the "received" view, lest they appear unscholarly. A telling illustration of this dubious stance is a lecture which Herbert Butterfield gave at Harvard in 1959. There he called attention to the process whereby the study of classics (Greek and Latin) was being replaced by the study of the history of science.[23] Of course, Harvard was not the place to be so specific about modern Western scientific culture as to recall its Christian and Catholic roots.

Butterfield, who was invited to Harvard on the strength of his book, *The Origin of Modern Science,* published four years earlier, would have broken the rules of academic etiquette and would have proved himself inconsistent, had he said something specifically Christian. For already in that book, whose first chapter had for title, "The Historical Importance of the Theory of Impetus," he shied away from pointing out the credal matrix of that historic breakthrough. His lack of candor (or courage) may have played a part in his taking, at the end of that chapter, as much praise away from Duhem as he had accorded to him. For, clearly, if the next chapter could have for title, "The Conservatism of Copernicus," then there was much more weight than what Butterfield cautiously granted to his statement, easily the most momentous statement in the entire book: "some historians of science have been disposed seriously to qualify the traditional concept of the 'Renaissance,' and to see, from the eleventh or twelfth century at least, a continuous development of Western thought."[24]

Four decades before Butterfield, Duhem noted the coming importance of the study of the history of science. He also displayed more balance and courage than Butterfield did. He praised at the conclusion of the third volume of his Leonardo studies the genius of Galileo for the discoveries he had made, though not for the ones he had merely borrowed from the medieval tradition, such as the inertial motion and the proportionality of space covered in accelerated motion to the square of time elapsed.[25] But Duhem, the historian of science, as long as he wanted to retain his intellectual integrity, could have no use for the myth, hatched in the 19th century, about the Renaissance as the dawn of human reason, and especially of scientific reason. Further, as a Christian, he had eyes for the most sinistrous aspect of that myth, namely, its virulent anti-medieval and, by implication, anti-Christian thrust. And as one most appreciative of the dignity of physics as an intellectual enterprise and of its central role in modern culture, he knew what was truly at stake in a true grasp of the nature and history of physics.

At stake is nothing less than the only effective intellectual resistance to the juggernaut of modern paganism moving ahead under a flag, the flag of science, which it is not entitled to fly, either logically or historically. To oppose that juggernaut is not only the sacred duty of Catholics, but also a basic condition for them for retaining cultural visibility in an age of science. One may bemoan the ever growing dependence of man on machines. One may deplore the threat posed by technology to the environment. One may look wistfully at bygone bucolic ages, but there is no turning the clock back. More science is needed to cope with problems created by science, although much more than science is also needed.

That additional commodity consists of such imponderables as intellectual perspective and moral resolve. Nothing about the latter can be supplied by science, some much publicized gurus notwithstanding. As to the intellectual perspective, that too is far more than a listing of facts and data. But in an age of science and an increasingly scientific age, the intellectual perspective in which science and its history are seen will be of paramount importance. The alternatives are few, in fact they are only two. Either science is seen as a supreme seal on man's autonomy or it will be seen as a gift from above.

Seen in that perspective the birth of science will show a startling parallel with that birth, the Incarnation, which is the greatest gift from above. If the latter had to come with as little stir as possible, the coming of science too, in terms of the

impetus theory, will appear natural in its unpretentious circumstances.[26] Buridan could give no news conference, he could not
expect that fanfare of publicity that goes nowadays with the
announcement of any major discovery. But in its historical
context Buridan's feat was part of that gigantic uncompromising
attitude which medieval Christianity took towards Antique
paganism as it resurfaced in the sweeping cosmological system
of Aristotle. Duhem saw this with a prophet's perspicacity and
courage. That so few even among his own have listened to him
is another telling vote on behalf of his status as one of the great
true prophets of modern times.

Notes

Chapter One

[1] See Text 9.

[2] This and other landscapes drawn by Duhem in the vicinity of Peyreleau are reproduced in the album, *The Physicist as Artist: The Landscapes of Pierre Duhem*, with my Introduction and Notes (Edinburgh: Scottish Academic Press, 1988).

[3] A. Rey, "La philosophie scientifique de M. Duhem," *Revue de métaphysique et de morale*, 12 (1904), pp. 699-744.

[4] *Le retour éternel et la philosophie de la physique* (Paris: Flammarion, 1927). Rey's efforts to reinstate to intellectual respectability this old pagan view of the universe are discussed in ch. 13 of my *Science and Creation: From Eternal Cycles to an Oscillating Universe* (Edinburgh: Scottish Academic Press, 1974), a work devoted to the entire history of that view from all major ancient cultures to the present times.

[5] "Physique et métaphysique," *Revue des questions scientifiques*, 34 (1893), pp. 55-83, also published in *Annales de philosophie chrétienne*, 127 (1893), pp. 461-86.

[6] A. Rey, *Le retour éternel*, p. 139.

[7] Further discussed in my books, *The Paradox of Olbers' Paradox* (New York: Herder and Herder, 1969) and *God and the Cosmologists* (Edinburgh: Scottish Academic Press, 1989).

[8] A remark of D. Parodi in his review of Charles Dunan's *Les deux idéalismes* (Paris: F. Alcan, 1911) in *Revue de métaphysique* 11 (1911), p. 240.

[9] The disappearance of this phrase, which graced the first edition of Rey's book, from its second, revised, edition (1923), was a telltale sign of the fact that, following the trials of World War I, Renan and scientism went into a brief eclipse in France.

[10] *Les Origines de la statique*, vol. 1, p. iv.

[11] C. Truesdell in his review (*Speculum* 36 [1961], p. 121) of M. Clagett's *The Science of Mechanics in the Middle Ages* (Madison: University of Wisconsins Press, 1961).

[12] *Travaux du Congrès International de Physique, Paris 1900*, ed. C.-E. Guillaume and L. Poincaré, vol. IV (Paris: Gauthier-Villars, 1901), p. 7.

[13] Maritain did so in his article, "La science moderne et la raison," *Revue de philosophie* 16 (1910), pp. 575-603. In the *Dictionnaire alphabétique et analogique de la langue française*, the French counterpart of the *Oxford Dictionary of the English Language*, Le Dantec is credited with the coining of the word *scientisme* (Part 7, R–Z, p. 370), with 1911 given as its first appearance.

[14] For details, see ch. 9 in my *Uneasy Genius: The Life and Work of Pierre Duhem*.

[15] "Usines et laboratoires," *Revue philomathique de Bordeaux et du Sud-Ouest*, Septembre 1899, pp. 385-400.

[16] See Text 15.

[17] It is revolutionary by subverting the basic thesis of the Enlightenment according to which Christian religion had to be discredited and discarded so that science, the true savior of mankind, may make its appearance. It is also revolutionary by its replacing the fictional or "Renaissance" origins of modern science with solid documentary information relating to the Middle Ages. The importance of *Les origines de la statique* is attested also by its being recently translated, almost a hundred years after its publication, into English by G. F. Leneaux, V. N. Vagliente, G. H. Wagener, with my Foreword, under the title: *The Origins of Statics: The Sources of Physical Theory* (Dordrecht, Boston, London: Kluwer Academic Publishers, 1991).

Chapter Two

[1] Hélène Pierre-Duhem, *Un savant français: Pierre Duhem* (Paris: Plon, 1936), p. 86.

[2] Ibid., p. 39.

[3] The Abbé V. L. Bernies in his reminiscences on Duhem, "Pierre Duhem," in *Revue des Jeunes* 15 (1917), p. 519.

[4] See Text 1. I wish to express my sincere thanks for the utilization of this priceless document to Mlle Marie-Madeleine Gallet, granddaughter of Albert Dufourcq, Duhem's colleague at the Faculty of Bordeaux and his best friend. Duhem's daughter, Hélène, received during her last ten years untiring support from Mlle Gallet who played a crucial part in securing Duhem's scientific correspondence for the Archives de l'Académie des Sciences.

[5] For details, see the doctoral dissertation by P. Brouzeng, "L'oeuvre scientifique de Pierre Duhem et sa contribution au développement de la thermodynamique des phénomènes irreversibles," (Université de Bordeaux, 1981), which, in good Marxist style, begins with the analysis of the workers'salary during Duhem's teaching career. With this in mind, one will not be misled by the title, *Science et Providence* (Paris: Belin, 1987), of Mr. Brouzeng's account, long in illustrations

(at times irrelevant) but short on genuine appreciation, of Duhem's life and work.

[6] See Text 2.

[7] *Un savant français*, p. 21.

[8] In the *Catalogue général des livres imprimés de la Bibliothèque Nationale* (vol. 13, pp. 35-36) a dozen publications (including a two-volume dissertation) by the abbé Biehler are listed from between 1879 and 1885. They all relate to higher mathematics, to the theory of equations and the properties of various curves. His work and life would deserve a special study which may perhaps provide interesting information about Duhem's life at Stanislas.

[9] See Text 2.

[10] *Un savant français*, p. 38. Récamier wrote his long letter about the same time when, at the urging of Hélène Duhem, two other reminiscences about Duhem were written by E. Artur, professor of English literature at the University of Lille, and by L. Marchis, first Duhem's student at Lille and later his colleague at Bordeaux. The soliciting of these letters, which greatly helped Hélène in writing the biography of her father and remain indispensable for any student of Duhem was but one of Hélène's many contributions to his father's memory. They will be set forth in detail, in my forthcoming book, *Reluctant Heroine: The Life and Work of Hélène Duhem*.

[11] *Un savant français*, p. 39.

[12] Particularly telling were in this respect Perrin's cursory references to Duhem in his *Traité de chimie physique. Les Principes* (Paris: Gauthier-Villars, 1903. Two years later, Perrin kept complete silence about Duhem as he discussed at a meeting of the Société française de philosophie the essential contents of the laws of thermodynamics. See its *Bulletin* 6 (1906), pp. 81-111.

[13] In view of the fact that Hadamard was a prominent liberal, in fact, the President of the Ligue Internationale du Droit d'Hommes, it is even more illogical to assume that he would not have noticed Duhem's antisemitism if it existed in Duhem in the sense of a hatred for someone just because he happened to be a Jew. He resented those Jews, and they were rather numerous, who were in the forefront of militantly anticlerical forces in the Third Republic. Duhem was too busy with his scientific work to oppose politically or socially anyone, Jew or not. He was, however, most eager to see as much of France's catholic cultural traditions maintained as possible. He was no less justified in this respect than anyone in present-day Israel supportive of Mosaic laws as forms of public legislation. At any rate, Duhem was not the kind of democrat for whom the momentary majority represented the supreme standard in thinking and comportment.

[14] P. Duhem, "Bronislas-Etienne Wasserzug," *Bulletin de l'Association des Anciens Elèves de l'Ecole Normale*, Tome 1884–89, pp. 57–62.

[15] J. Hadamard, "L'Oeuvre de Duhem dans son aspect mathematique," in *L'Oeuvre scientifique de Pierre Duhem* (Paris: Blanchard, 1928), p. 467.

[16] *Le Temps* (Supplément littéraire), August 11, 1936.

[17] E. Picard, "La vie et l'oeuvre de Pierre Duhem," Document Institut 1921–34. (Paris: Gauthier-Villars, 1921), p. 4.

Chapter Three

[1] See L. Trenard, *De Douai à Lille, une université et son histoire* (Lille: Université de Lille III, 1978), p. 84.

[2] These words must have often been repeated among the Duhems. Hélène Duhem quoted them three times in her *Un savant français*, pp. 26, 53, 146.

[3] Dossier Duhem, p. 125. This is my pagination. In that Dossier the documents are placed in the reverse chronological order.

[4] All these essays, published by Duhem during his years in Lille, were reprinted as a single volume, with my introduction, under the title, *Prémices philosophiques* (Leiden: E. J. Brill, 1987). Excerpts from three of them are given in Texts 3-6.

[5] *Un savant français*, p. 87.

[6] A photocopy of that letter was part of the display on Duhem, at the College Stanislas, in early December, 1988.

[7] *Un savant français*, p. 96.

[8] Ibid., p. 98.

[9] *Rapport général de l'Université de Bordeaux 1893-94*, p. 15.

[10] Archives de l'Université de Bordeaux.

[11] Dossier Duhem, pp. 104-5.

[12] Both these letters are part of Duhem's scientific correspondence, now in the Archives de l'Académie des Sciences, Paris.

[13] Dossier Duhem, p. 86.

[14] Text 7.

[15] Dossier Duhem, p. 56.

[16] "Thermochimie, à propos d'un livre récent de M. Marcelin Berthelot," *Revue des questions scientifiques* 42 (1897), p. 376.

[17] Letter of July 3, 1901, to Liard. Quoted in full in *Un savant français*, pp. 119-24.

[18] For details, see my *Uneasy Genius*, p. 335.

[19] *Un savant français*, p. 26.

[20] "Thermochimie . . . ," p. 392. *Imitation du Christ*, Book I, ch. 3, par. 5.

[21] "Paul Tannery 1843–1904," *Revue de philosophie* 4 (1905), p. 229.

[22] Dossier Duhem, p. 85.

[23] Archives de l'Université de Bordeaux.

[24] Dossier Duhem, p. 20.

[25] *Un savant français*, p. 170.

Chapter Four

[1] *Un savant français*, p. 150.

[2] *Zeitschrift für Physikalische Chemie* 19 (1896), p. 518.

[3] Ibid., 30 (1899), p. 183.

[4] From Poincaré's Preface to his *Thermodynamique. Leçons professées pendant le premier semestre 1888-89*, rédigées par J. Blondin (Paris: Georges Carré, 1892), p. xix.

[5] *Bulletin des sciences mathématiques* 17 (1893), p. 15.

[6] C. Truesdell, *Rational Thermodynamics* (New York: McGraw-Hill, 1969), p. 35.

[7] H. G. Bryan, "Pierre Duhem," *Nature* 98 (1916), p. 132.

[8] The letter, of January 11, 1902, was first published by P. Brouzeng, "Magnétisme et énergétique. La méthode de Duhem. A propos d'une lettre inédite de Pierre Curie," *Revue d'histoire des sciences* 31 (1978), pp. 333–44.

[9] See *Uneasy Genius*, p. 309

[10] Reported by R. Resnick, a participant at that conference, in his essay, "Misconceptions about Einstein: His Work and his Views," *Journal of Chemical Education* 52 (1980), p. 860.

[11] *Notice sur les titres et travaux scientifiques de Pierre Duhem* (Bordeaux: Gounouilhou, 1913), p. 107.

[12] "L'Ecole anglaise et les théories physique," *Revue des questions scientifiques* 34 (1893), p. 367.

[13] *Notice*, p. 107.

[14] "Quelques réflexions au sujet de la physique expérimentale," *Revue des questions scientifiques*, 36 (1894), p. 207. See Text 4.

[15] He did so first in an essay published under that title in 1932 which later became the first chapter in his book, *Le réalisme méthodique* (Paris: Téqui, 1936). Translated into English bu P. Trower, *Methodical Realism*, with an introduction by this author (Front Royal, Va.: Christendom Press, 1989).

[16] The question is discussed in detail in my *God and the Cosmologists* (Edinburgh: Scottish Academic Press, 1989), based on eight lectures given at Oxford in November 1988.

[17] "Physique de croyant," in *La théorie physique: son objet et sa structure* (2d ed.: Paris: Marcel Rivière, 1914), p. 471. See Text 9.

[18] Part of this letter was preserved through E. Picard's ample quotations from it in his éloge of Duhem, "La vie et l'oeuvre de Pierre

Duhem," Document Institut 1921–34 (Paris: Gauthier-Villars, 1922), p. 40.

[19] See Text 21.

[20] E. Renan, *Souvenirs d'enfance et de jeunesse* (1883; Paris: Calmann Levy, 1902), p. 251.

[21] H. Metzger, *Les concepts scientifiques* (Paris: F. Alcan, 1926), p. 160.

[22] For further discussion, see ch. 15, "Paradigms or Paradigm," in my Gifford Lectures, *The Road of Science and the Ways to God* (Edinburgh: Scottish Academic Press, 1978).

[23] See Text 8.

[24] See Texts 10, 11, 12, 18, 24.

[25] *Le Système du monde* (Paris: Hermann: 1913–59), vol. IV, p. 453.

[26] Ibid. See Text 20.

[27] P. Tannery, *La géometrie grecque* (Paris: Gauthier-Villars, 1887), pp. 10–11.

[28] *Le Système du monde*, vol. II, p. 390. See Text 19.

[29] For a preliminary report, see my article, "Science and Censorship: Hélène Duhem and the Publication of the *Système du monde*," in *The Intercollegiate Review* vol. 21, Nr 2, Winter 1985–86, pp. 41–49. The full story will be given in my *Reluctant Heroine: The Life and Work of Hélène Duhem*.

[30] *Le Système du monde* , vol. VIII, pp. 329 and 331. Advance glimpses of this were given by Duhem already in his *Etudes sur Léonard de Vinci: Ceux qu'il a lus et ceux qui l'ont lu* (Paris: Hermann: 1906–13), vol. III, p. 52.

[31] For further discussion, see my books, *Science and Creation: From Eternal Cycles to an Oscillating Universe* (1974; 2d enlarged ed.; Edinburgh: Scottish Academic Press, 1986) and *The Savior of Science* (Washington, D.C.: Regnery-Gateway, 1988).

[32] *The New York Times*, Dec. 21, 1983, p. A26, cols. 4-5.

[33] E. Renan, *L'Avenir de la science* (1890; Paris: Calmann Lévy, n. d.), pp. 312-13.

[34] E. Renan, *Dialogues et fragments philosophiques* (1876; 5th ed.; Paris: Calmann Lévy, 1903), p. 178.

[35] E. Renan, *Feuilles détachées* (1892; 19th ed.; Paris: Calmann Lévy, 1924), pp. 156-57.

[36] E. Renan, *Dialogues et fragments philosophiques,* p. 288.

[37] E. Renan, *L'Avenir de la science,* p. 37.

Chapter Five

[1] See *Uneasy Genius,* p. 173.

[2] Quoted in V. Giraud, *Essai sur Taine, son oeuvre, et son influence d'après des documents inédits* (2d ed.: Paris: Hachette, 1901), p. 75.

[3] A point, however elementary and crucial, which is invariably glossed over in various articles on Duhem by R. H. D. Martin.

[4] *Un savant français,* p. 140.

[5] Ibid., p. 127.

[6] Duhem emphatically agreed with Taine's blaming the ideology of the French Revolution for the anarchistic tendencies in late 19th-century France. See my *Uneasy Genius*, p. 154.

[7] *Un savant français*, p. 181. There one finds the original account of several other moving examples of Duhem's generous compassion toward the poor and the sick, summarized in the following paragraphs.

[8] Ibid., p. 179.

[9] From the eulogy of Duhem, delivered by the Abbé Bergereau on November 19, 1916, in the Chapelle de la Madeleine, Bordeaux, printed in the brochure, "Pierre Duhem . . . membre fondateur de l'Association Catholique des Etudiants de l'Université de Bordeaux" (Bordeaux: Wetterwald Frères, 1916), pp. 6–8.

[10] Duhem's speech was printed in the *Compte rendu annuel 1913-1914* of the Association (Bordeaux, 1914), pp. 39–44.

[11] See Text 27.

[12] See Text 26.

[13] Quoted from Duhem's obituary by E. Jordan in *Annuaire de l'Association Amicale de Secours des Anciens Elèves de l'Ecole Normale Supérieure* (Paris, 1917), p. 172.

[14] Ibid.

[15] Quoted by the Abbé L. Garzend, "In Memoriam P. Duhem," *Cahiers catholiques*, Feb. 10, 1922, p. 1085.

[16] This is one of the almost daily letters, many of them but a couple of phrases, which Duhem wrote to his daughter during the academic year between 1909 and 1916.

[17] Both these drawings are reproduced in the album, *The Physicist as Artist*, referred to in Chapter 1, note 2.

Chapter Six

[1] It is given in full in Text 15. This letter survived only in its being printed in *Un savant français*, pp. 158–69. Obviously, Duhem kept a copy of the letter in view of the importance he had attached to it. My search for the original in the Archives of the Institut Catholique was fruitless.

[2] *Compte rendu du Troisième Congrès Scientifique International des Catholiques tenu à Bruxelles du 3 au 8 Septembre 1894. Troisième Section . . . Sciences philosophiques* (Bruxelles: Société Belge de Librairie, 1895), pp. 323–24.

[3] *Un savant français*, pp. 157–8.

[4] For an analysis of their rapport and exchange of letters,, see my article, "Le Physicien et le metaphysicien: La correspondance entre Pierre Duhem et Reginald Garrigou-Lagrange," *Actes de l'Académie Nationale des Sciences, Belles-lettres et Arts de Bordeaux* 12 (1987), pp. 93–116.

[5] See Text 17.

[6] A. de Lapparent, *Science et apologétique* (12th ed.; Paris: Bloud et Gay, 1914).

[7] L. de Launay, *L'Eglise et la science* (Paris: Bernard Grasset, 1936).

[8] A far greater audience did Dufourcq reach through his essay, "Les origines de la science modernes d'après les découvertes récentes," *Revue des deux mondes* 16 (1913), pp. 349–78.

[9] *L'Apologétique. Nos raisons de croire, réponses aux objections,* publiée sous la direction de M. Brillant et M. Nedoncelle (Paris: Bloud & Gay, 1948), p. 604.

[10] Indeed its half a dozen pages (10-16) on "Science in the Middle Ages" were written as if Duhem had never existed.

[11] The eight volumes of L. Thorndike's *History of Magic and Experimental Science* were published between 1923 and 1958 (New York: Columbia University Press).

[12] See, for instance, his widely read *Progress and Religion,* first published in 1936.

[13] A. N. Whitehead, *Science and the Modern World* (New York: Macmillan, 1926), p. 11.

[14] Ibid.

[15] It is revealing that the second reference in Koyré's *Etudes galiléennes* is to Bachelard, the first to propose, in his Le *Nouvel esprit scientifique* (1934), the idea of such mental mutations as an explanation of intellectual history.

[16] For a discussion of the "religious" background of Koyré's interpretation of the history of science, see ch. 15, "Paradigms or Paradigm," in my Gifford Lectures, *The Road of Science and the Ways to God.*

[17] See ch. 7, "Medieval Science in the Christian West," in vol. II, *History of Science. Ancient and Medieval Science,* ed. R. Taton, tr. A. J. Pomerans (New York: Basic Books, 1963), pp. 468-532.

[18] For the copy of De Broglie's letter, see Archives de l'Académie des Sciences. See also note 10 to ch. 4.

[19] D. G. Miller, "Ignored Intellect: Pierre Duhem," *Physics Today* 19 (December 1966), pp. 47-53. In French, "Pierre Duhem, un oublié," *Revue des questions scientifiques* 28 (1967), pp. 445-70.

[20] In more than one sense. In a friendly conversation one of them voiced his surprise that there was a Commune in Paris in 1871!

[21] The title of that story will be "The Duhem Scandal," as neglect of Duhem among Catholic intellectuals in France and outside it is nothing short of being scandalous.

[22] A. Maier, *Die Vorläufer Galileis im 14. Jahrhundert* (Rome: Edizioni di Storia et Letteratura, 1949), p. 1. See in this connection a remark about my *Uneasy Genius* in its review in *Archives de philosophie* 49 (1986), pp. 338–40.

[23] H. Butterfield, "The History of Science and the Study of History," *Harvard Library Bulletin* 13 (1959), pp. 329–47.

[24] H. Butterfield, *The Origins of Modern Science 1300-1800* (1949; New York: Macmillan, 1960).

[25] "En faveur de ces lois, Galilée pourra bien apporter de nouveaux arguments, tirés soit du raisonnement, soit de l'expérience; mais, du moins, il n'aura pas à les inventer." *Etudes sur Léonard de Vinci*, vol. 3, p. 562.

[26] *The Savior of Science* (Washington, D.C.: Regnery Gateway; Edinburgh: Scottish Academic Press, 1988).

TEXTS

In the identification of the French original of the subsequent Texts, numbers in bold refer to the entry in the list of Duhem's publications in my book, *Uneasy Genius: The Life and Work of Pierre Duhem* (1984; 2nd enlarged paperback edition: Dordrecht, London, Boston: Martinus Nijhoff, 1987), pp. 437–456.

1

The Prussians at Chateaudun

Letter of November 1870 about the siege of Chateaudun*

The siege of Chateaudun

On Tuesday, the 18th of October, we sat down as usual at table and after lunch our dear maid went to the market and returned almost immediately, saying that the Prussians had just arrived. For some time there had been many false alerts. Last week [the villages of] Varise and Lutz were set on fire, but in spite of this we were not afraid. Then came my cousin, who was the attorney of the town, followed by a volunteer (the quarter-master Chancerel) who lodged with us, saying that the Prussians, 6000 strong, with 24 cannons, 2 heavy guns, and 1 machine gun, were at 600 meters from the town and that we had to take refuge in the hospital which probably would be respected. At the gate we heard the first cannon-burst (it was half past twelve). In the hospital we were confined (after having embraced our dear papa who had just seen the Prussians line up for battle) in the cold and humid cellar together with the white-clad sisters, their pupils, and perhaps 300 women and children.

*Translated from a photocopy of the original made available by Mlle Marie-Madeleine Gallet.

The defenders of the town consisted of 900 volunteer sharp-shooters from the Seine (Colonel Lippouski), of 100 to 150 volunteer sharpshooters from Nice and Nantes, and 300 to 400 national guardsmen (Commander Testanière) who fought ferociously until six in the evening.

We remained standing against the pillars of the cellar for nine hours, in the midst of the sick who were hit by bullets in their beds, brought downstairs half naked, and operated upon there. From time to time the sisters passed by, escorting with small funeral lamps the dead who were laid down in one of the caverns. The cannons, the shooting and our prayers could alone be heard; my mother could not help thinking of the last act of the Huguenots, because we were in the same position. Night came and we hoped that with it the horrible slaughter would end, but far from it, the cold got hold of us more and more, we did not even have a piece of bread and only the terrible fire gave us some light in the cellar. Exhausted by fatigue, my sisters and I slept on a big heap of firewood. Then at nine the sisters came to tell us that they could no longer be responsible for us and that the conflagration was gaining ground, and that the Prussians had broken through the barricades and were in the town and could come to butcher us in the cellar. It was then that there was a rush to the doors where, frightened by the fire and the shooting, everyone pulled back. One lady, my mother and three of us children alone had the courage to exit; nobody wanted to follow us; we wanted to find papa at any cost and, if necessary, die with him.

Having reached the street, the sentinels of the sharpshooters, whom we met from barricade to barricade, did not want to let us go, telling us that we were running into our death; in fact it was then that the battle on the main square was at its height.

From the upper end of the main street we caught a terrifying sight; in front of us the square looked like a red-hot brazier, above the fountain there was nothing but a shower of bullets and big shells launched by the cannons and the machine guns.

Thanks to our imploring the head of the patrol, we obtained permission to go on and arrived to the house where papa and my cousin received us with open arms. Mama fell down in a faint; she was at the end of her strength, but our weariness was far from over.

Our departure from Chateaudun

At half past ten the president of the Court came to rejoin us, bringing a small bag; my father and my cousin each prepared one, containing our money, and these gentlemen went out in order to ascertain if our flight was possible, unable to be of further service to us. My father found some poor women who told him that the Prussians were setting fire with their own hands to the houses of the main square, that we should flee as quickly as possible, that we were alone because everyone else had left. Then mama, who put us down to sleep fully dressed on mattresses in the salon, woke us up and we all took to the road to St Jean at 11 at night.

An enormous barricade blocked the road, we clambered over it on all fours, helped by the president and my cousin.

At half past midnight we arrived at the Bordas farm, where we rested while waiting for the day; the cannons boomed time and again and the conflagration kept growing. We could not refrain from turning all the time to view that horrendous spectacle.

At five o'clock in the morning we started out for Legron, whence a peasant woman led us in her small carriage to Goury and from there to Brou where we hoped to find rest. But no sooner had we finished our lunch than they began shouting that the Prussians had arrived; everyone cried and sought safety, and in spite of our fatigue we were soon on the march again. It was only after more than an hour of walking during which I had to carry my sisters more than once on my back, that we found a big carriage full of refugees who, in spite of their own load, kindly took us on; 22 of us were sitting on sacks of oats.

It rained heavily, we were dying of hunger, of cold and of fatigue when we arrived at Beaumont where we spent the night in an inn. Next morning we started out for Nogent le Rotrou where after having said good-bye to my cousin and the president, we climbed into a third-class railroad carriage and after 18 hours on hard benches we arrived in Bordeaux at nine in the evening; we had not changed our clothes for three days.

P. Duhem

Loss of volunteers   100 men
Loss of Prussians   2500 men

2

Collège Stanislas

"Souvenirs de l'Ecole préparatoire (1878-1882)," in *Centenaire du Collège Stanislas (1804-1905)* (Paris: Imprimerie de J. Dumoulin, 1905), pp. 101-122 (**1905-22**).

It was bursting with the ardor of life, around 1880, that courtyard reserved for the recreations of the Preparatory School. Candidates for the Ecole Polytechnique or for Saint Cyr, aspirants to the Ecole Centrale or to the School of Forestry mingled there joyfully. From one class to the other there was much bantering, for each was — or at least pretended to be — convinced that he was preparing for the only profession worthy of an honest man's ambitions. But basically we all knew that no army can consist of artillery alone, or just of cavalry, or of foot-soldiers for that matter, that a nation had equal need of officers, of engineers, of professors. We all knew that we all had to collaborate for the well-being and prosperity of France. Beneath a superficial rivalry were also formed the ties of solid friendship capable of withstanding long separations. How ready to be reborn is that friendship, warm and joyful, when around the corner of a street, in an unknown town where, exiled by the happenstance of an appointment, one encountered, under the four-ribboned cap of a colonel of the artillery or of the hussars, a friend whom one had not seen for twenty-five years! How the hands moved and clasped as one recalled memories, already very old, of passionate battles where, lined up on opposite sides of the yard, "mice" and "little horns" bombarded one another valiantly with snowballs! . . .

Among the countless memories that rise in my heart and fill it, among the venerable figures who come to life in the eyes of my imagination, let me be allowed to make a choice. Let me evoke especially those teachers who guided us to the doors of the Ecole Polytechnique or of the Ecole Normale Supérieure. . . .

After a careful initiation into elementary mathematics, instruction in advanced mathematics seemed to approach its highest point. Of course, the classes were taught by Vazeille. Since the time when, on the benches of Stanislas I followed the discussion of equations in S, I have attended many lectures. Some of them I have appraised with the meticulous curiosity of the specialist; some others I have followed with the attention of a fervent disciple; none of them have evoked in me the feeling of perfection to the degree as did Vazeille's teaching.

I still seem to hear and see that admirable teacher: On a body lean and erect, a head with a profile of remarkable purity that was framed by a straight beard and ample tresses; an expression somewhat exalted which a fine smile sweetened now and then; a diction of impeccable correctness, pronounced by a warm and engaging voice; a fine hand which arranges, on a long blackboard, in perfect order equations that look like calligraphy or traces, with extraordinary sureness, the connections of a complicated diagram.

The word *elegant* was one of those that Vazeille pronounced readily. It certainly best characterized his teaching. His course was a true work of art. Each of the chapters which composed it was lovingly chiseled. In their turn the algebraic method and the geometric method used by him seemed to rival one another in power and skill. This emulation between the two procedures by which the human mind takes cognizance of the mathematical truths let the theories unfold in a perfectly balanced symmetry which barred all monotony. No affectation whatever in that elegance! The absolute clarity, the irreproachable ordering of the theories, which Vazeille set forth, had their raison d'être in the very nature of the problems treated, in the penetrating intuition by which the professor grasped their nature. No artifical simplifications, not one of those too easy procedures which success alone justifies, was allowed in the advanced calculus classes at Stanislas.

With no letup Vazeille asserted that the general method is always the most direct, the shortest, and simplest, provided one knew how to make use of it. The ease with which he gave the solution of the most difficult problems more than amply proved the justness of the principle of which he made himself the champion. . . .

One cannot think of a contrast more gripping than the one between Vazeille and our professor of physics and chemistry, Moutier. All the "mice" of Paris knew this big body, this powerful amplitude, this large face, this disheveled bear, these eyes resembling two goggles, all this mobile physiognomy which delighted in putting on the funniest and most unexpected expressions.

If the class taught by Vazeille resembled a salon, the one taught by Moutier was almost as vibrant as a hall at the hour of public auction. The professor did not feel at ease except when in the middle of that exuberant, tumultuous, noisy life, of that banter that went from bench to bench, of that stamping of the feet that shook the amphitheater. On occasion, though very rarely, we would complain, or we thought to complain, to our teacher; then we would come up with a terrible conspiracy by decreeing a silent classroom! On those days, after we had in vain tried to give free rein to our captive tongues and to unfreeze our rigid miens, Moutier would try to go on with his course but he would not succeed. He would get mixed up in his reasonings, lose track of his calculations, and end by throwing down the chalk in a rage. Then, with our vengeance taken, we would give free rein to our tumultuousness; immediately our teacher's face would become serene; he would resume his interrupted calculations and, smiling in the midst of a deafening bedlam, he would complete them.

Delivered before an audience at the boiling point, his course was marvelous in its neatness, precision, conciseness. Not a demonstration that had not been reduced to absolutely needed propositions. Not a law whose enunciation had not taken the form of absolute rigor. A few words of extreme soberness, sufficient to mark the hypothesis to be taken with caution, the experimental procedure of doubtful value, to form the critical sense of his students, such was the aim which Moutier set himself and it would

have been difficult to attain it more exactly than he did. . . . Yet
Moutier was, in certain parts of physics an audacious and powerful
inventor. Familiar with the laboratories of Henri Sainte-Claire
Deville, he witnessed the birth of the theory of chemical dissociation. He recognized right away that thermodynamics alone could
organize that novel domain. In the name of that science he took
possession of it before anyone else. While Hortsmann in Germany
and Willard Gibbs in the United States have marvelously developed the truth which he was the first to enunciate, Moutier
discovered the true sense which it was proper to attribute, in
physical chemistry, to the sign of heat developed in the course of
a reaction. His discovery put an end to the hesitations that for a
century guided a Lavoisier, a Laplace, a Berthollet, a Favre, a
Thomsen, and a Berthelot. The course of chemistry at College
Stanislas was given the first fruits of that great principle of natural
philosophy. . . .

One day I was called for questioning in physics by an
instructor to whom none of us had until then to render account.
He was a priest, very young, whom his blonde hair and clear face
made look even younger. He had just left his dear Bretagne in
order to prepare in Paris for his *agrégation* in mathematics which he
soon was to take on in grand style. He was not altogether
unknown to us. A few years earlier as a student at the Ecole des
Carmes he had followed courses in advanced mathematics at
Stanislas. The vivacity of his spirit, his aptitude for the exact
sciences, no less than his adroitness and sincere friendliness, left
vivid memories — almost a legend — in the Preparatory School.
During that first "drilling" he did not interrogate me much. He
spoke rather of the teachers he was glad to meet again, of the
house to which he was happy to return. On that day there was
born between us a friendship which the passing of time could but
make stronger and more trusting. That young Breton priest did
not then suspect that twenty years later the teachers of whom we
spoke with affection would take the road to exile after having
been exiled from the very house which they made grow and
prosper. He did not foresee that the Jacobin spirit would blow like
a tempest and that, in order to govern Stanislas in those torments,

to pilot it in the midst of reefs, the alumni of Stanislas would turn to his firm hand and farsighted views.

Nobody, for sure, foresaw the storm as the Preparatory School flourished under the pleasant direction of the Abbé Biehler. His figure calm and serene, under the invariable velvet cap, his face smiling behind the immovable glasses, our director chatted with a few "mice" in a corner of the vast courtyard. The conversation centered on some advanced theorems of algebra or calculus. He had, the previous day, presented its first principles in one of his conferences where the always same voice, the always impeccable diction set forth the proofs with extreme rigor and perfect elegance. Now he extended his lecture into a formal conversation by showing us, behind the propositions already familiar to us, an unlimited domain of new truths which made shine before our eyes, behind familiar lights, the infinite splendor of science.

Some utilitarians distanced themselves from the group that surrounded M. Biehler. They disdained these high speculations. What is the good of them if not inscribed in the syllabus of any school and if no examiner ever inquired about them? But these utilitarians were few in number at that time and certainly not respected at Stanislas. We made it a matter of honor to pursue with passion the pure truth, for its own sake, for the love of its beauty. We spurred ourselves to disdain the dishonest calculations of those who with the program of competitive exams under their arms did business with science and wanted to impose on it a tariff. We would not accept success bought at reduced price and, as it were, at a discount.

At any rate, that religious, dressed in black, with an expression timid and sweet, commanded our respect and our admiration. We knew him to be capable of taking his place among the best theoreticians, next to an Appell, a Picard, a Poincaré, whose first publications were then received with favor by leading mathematicians. His thesis on the theory of elliptical functions — a true masterpiece of algebraic elegance — provoked the Sorbonne's applause whose echo carried as far as us. We knew that only the religious vocation of teaching, the desire to form men — Frenchmen and Christians — had alone prompted him to renounce science, its joys and honors. We guessed how much that sacrifice cost him and we felt our heart moved with gratitude.

From time to time a limping old man, leaning on his huge cane, in pain and out of breath, slowly ascended the staircase that led to Mr Biehler's small room. On occasion a new student watched with astonishment that laborious ascent. But the smile that already appeared on his lips, quickly turned into a respectful curiosity when an older student whispered into his ears the name of that handicapped old man: Hermite! The great mathematician had just left his working retreat for a few hours. He came to visit a religious who had been his pupil and who became his friend.

Hermite could not have given a greater proof of his affection toward Mr Biehler than to inquire about his dear "mice" and to quiz some of them. Overcome by the closeness of the one whom the entire scientific Europe viewed as the purest incarnation of the mathematical spirit, the student did his best to discuss the theories which were taught to him. Then, not without astonishment and confusion he heard the great mathematician make ecstatic exclamations about the beauty of proofs recited to him and even the talent for algebra of the one who did the recitation. When the latter returned to the study hall after one such quiz, he had the awareness of having seen a genius at close range.

On occasion also a tall priest, his body emaciated, his back bent, his looks tired and colored by a ravaging sickness, went across our courtyard at slow pace. He was soon to die a cruel death. Everybody knew it and he knew it better than everybody. He walked along, surrounded by profound respect, and controlling his pains he smiled at us benignly. He knew each of us thoroughly as a father knows his children. He spent on us with a surprising and supreme solicitude all that the ravaging sickness left to him in the way of strength and activity. And after the Abbé de Lagarde had spoken to us, we were left troubled to the depth of our souls because we felt that we have just talked with a saint!

3

The Perfection of Science

From "L'Ecole anglaise et les théories physiques," *Revue des questions scientifiques*, 34 (1893), pp. 372-73 (**1893-9**).

Ideas now spreading among physicists pose a serious problem which can be formulated as follows: Is incoherence in logic legitimate in physical theory? or, . . . Is it legitimate to symbolize either several distinct groups of experimental laws, or even one single group of such laws, by several theories of which each rests on hypotheses that cannot be reconciled with the ones that bear on the others?

To this question we do not hesitate to reply in the manner in which we have already done: IF ONE OMITS TO INVOKE THE REASONS OF PURE LOGIC *one cannot prevent a physicist from representing by several mutually irreconcilable theories either different ensembles of laws, or one single group of laws; one cannot* [in that case] *condemn incoherence in the development of physical theory.*

If one admits, in fact, as we have tried to establish, that a physical theory is nothing but a classification of an ensemble of experimental laws, how could one derive from the code of logic the right to condemn a physicist who would use, in order to coordinate different ensembles of laws, procedures of different classifications, or a physicist who would propose, for one and the same ensemble of laws, different classifications derived from different methods? Does logic prohibit naturalists to classify a group of animals according to the nervous system and another group according to the circulatory system?

Would a malacologist proceed absurdly if he sets forth both the classification of Bouvier, who groups the molluscs according to the arrangement of nervous filaments, and the classification of Remy Perrier who bases his comparisons on Bojanus' study of the organs? Thus a physicist would logically have the right to consider matter as continuous and to take it at the same time to be formed of separate atoms; to explain capillary effects by attractive forces among immobile particles and to endow the same particles with rapid motion in order to account for heat effects. All these separate procedures would not violate the principles of logic.

Logic itself imposes on the physicist but one single obligation. The obligation is not to confuse these different procedures of classification with one another; thus when the physicist establishes between two laws a certain connection, he has to state precisely the one among the methods used by him that justifies that connection. Or, in a word, to use Poincaré's expression, not to mix two contradictory theories.

Therefore, if we invoke solely reasons of logical nature, we cannot condemn incoherence in physical theory. But reasons of purely logical nature are not the only ones that reasonably govern our judgments. The principle of contradiction is not the only one to which we have recourse. That we might legitimately reject one method, it is not necessary that it should be absurd. It is enough that by rejecting it we aim at a method more perfect. It is in virtue of this principle that we can cut through the difficulty we are examining and legitimately state the rule: *In physical theory we must avoid logical incoherence* BECAUSE IT HARMS THE PERFECTION OF SCIENCE.

4

No *experimentum crucis* in Physics.

From "Quelques réflexions au sujet de la physique expérimentale," *Revue des questions scientifiques* 36 (1894), pp. 196-99 (**1894-5**).

Here we touch on one of the essential points of the experimental method used in physics.

The reduction *ad absurdum*, which seems to be but a means of refutation, can become a method of demonstration. In order to demonstrate that a proposition is true, it is enough to trace its contradictory proposition to an absurd consequence. Well known is the profit which geometers have drawn from this kind of proof.

Those who assimilate the experimental contradiction to the reduction *ad absurdum* think that it is possible to follow in physics a method similar to the one of which Euclid made use in geometry. Do you wish to obtain for a group of phenomena a certain and incontestable theoretical explanation? List all the hypotheses that can be made in order to account for this group of phenomena. Then, by experimental contradiction, eliminate all except one. This latter ceases to be a hypothesis in order to become a certainty. Assume, in particular, that only two hypotheses are on hand. Look for experimental conditions such that one of the hypotheses predicts the production of the phenomenon and the other the production of a phenomenon altogether different. Realize these conditions and observe what takes place. According to whether you observe the first phenomenon predicted or the second, you discard the second hypothesis or the first. The one that will not be discarded will henceforth be incontestable; the debate is over, a new truth will accrue to physics. — Such is the *experimentum crucis*.

Two hypotheses are on hand concerning the nature of light. For Newton, Laplace, and Biot, light consists of projectiles launched with very high velocity. For Huygens, Young, and Fresnel light consists of vibrations whose waves propagate in an elastic medium. These two hypotheses are the only ones whose possibility is assumed. Either the motion is transported by the body which it drives or it passes from one body to another. Follow the first hypothesis. It tells you that light moves faster in water than in air. Follow the second. It tells you that light moves faster in air than in water. Set up Foucault's apparatus and put in motion the rotating mirror. Two luminous spots will form, one white, the other greenish. Is the greenish band to the left of the whitish band? If so, light moves slower in water than in air. This means the discarding of the undulatory hypothesis. Is the greenish band to the right of the whitish band? If so, light moves faster in water than in air. This means that the corpuscular hypothesis is discarded. You compare the position of the two bands. You see that the greenish band is to the right of the whitish band. The debate is settled. Light is not particle but an undulatory motion whose waves propagate in an elastic medium. The corpuscular hypothesis has its life terminated. The undulatory hypothesis ceased to be doubtful. It is now a new article of the scientific creed.

What we have said in the preceding paragraph shows how one is deceived by attributing to Foucault's experiment a meaning so simple and a bearing so decisive. It is not between two hypotheses, the corpuscular and undulatory, that Foucault's experiment decides but between two theoretical ensembles, each taken as a whole, between two systems, the optics of Newton and the optics of Huygens.

But let us admit for a moment that in each of these two systems everything is needed, everything is necessary, in terms of logical rigor, save one hypothesis. Let us therefore admit that the facts directly eliminate the only doubtful assumption which it includes. Does it then follow that one can find in the *experimentum crucis* an irrefutable means of transforming into certain truth one of the two hypotheses on hand, in the same way that the reduction *ad absurdum* of one hypothesis assures the truth of the contradictory theorem? Between two contradictory propositions of geometry there is no room for a third opinion. If one is false the other is necessarily true. Is it the same with two hypotheses of physics? Do we dare never to assert that no other hypothesis is imaginable? Light can be a stream of projectiles. It can be a vibratory movement whose waves are propagated by an elastic medium. Can it be but one or the other of these two? Arago may have thought so but it is difficult for us to share his conviction since Maxwell proposed to attribute light to periodic electric currents transmitted in a dielectric medium.

The experimental method cannot transform a hypothesis of physics into an incontestable verity because one can never be sure of having exhausted all imaginable hypotheses bearing on a group of phenomena. The *experimentum crucis* is impossible. The truth of a physical theory is not decided by heads or tails.

5

Between Scepticism and Positivism

From "Physique et métaphysique," *Revue des questions scientifiques* 34 (1893), pp. 71–73 (**1893–8**).

We have just set forth our essential thesis about the mutual independence of physical theories and metaphysical investigations. Let us now try to dissipate some of the objections that are most often made to that thesis.

Does one open the door to scepticism by insisting on the natural separation between physical theories and metaphysical investigations? Is this to make a concession to positivism?

It is almost impossible to mark the just limits of a science, the limits set to it both by the nature of the objects which it studies and by the nature of our intellect, without being charged right away with scepticism. It seems to some each one of the logical methods available to our reason is omnipotent and that each can bear on all subjects and reveal the most hidden secrets. In the workshop of human reason each tool is, according to them, appropriate to most diverse needs and our intelligence resembles somewhat the chemist who wanted to know how to file with a saw and how to saw with a file. Sad pretentions of dogmatism generate the worst errors and furnish scepticism with its most troublesome proofs. Ask a soul who is tormented by doubt, not by the facile doubt born of laziness and vanity but by the anxious and painful doubt arising from analysis and meditation. Search for the route along which doubt has entered that soul. Ask that soul how did his faith in reason vanish. You always receive a similar reply. The soul despaired because carefully connected deductions led it to a manifestly false conclusion, because a hard research refused to produce an expected result. Examine then where did that error, that sterility, come from: always from an unjustified extension given to a legitimate method of logic. The tool was set for a determined work, but the workman wanted to give it a another function. In vain did he handle it for a long time, display his skill; it either produced nothing or served only a wrong need. Thus rebutted, he threw away the tool and folded his arms.

Do you wish to guide that discouraged man back to work? Do you want him in the future to avoid miscalculations and deceptions? Teach him the exact usage of his tools. Teach him that a saw is not good but for sawing and file for filing. It is the same with the means of knowing that God has put at the disposition of our reason. Nothing is so effective in favoring scepticism as to confuse the domains of various sciences. Nothing, on the contrary, is more effective against this dissolving trend as the exact definition of the various methods and the precise tracing out of the field which each has to explore.

By denying to metaphysics the right to govern the researches of physics, by denying to the physical theories the right to pose as metaphysical explanations, do we become thereby positivists? We maintain that the positive sciences should be used with methods appropriate to positive sciences. We maintain that these methods rest on self-evident principles and can function independently of all metaphysical research. We maintain that these methods, efficacious in the observation of phenomena and in the discovery of laws, are incapable of grasping the causes and touch on the substances. But this is not to be a positivist. To be a positivist is to assert that there is no other logical method than the method of positive sciences, that what is inaccessible to that method, what is unknowable by positives sciences, is in itself and absolutely unknowable. Is this what we maintain?

Do you want to play the game of positivism? Then confuse the domain of metaphsyics with the domain of physics, the method of metaphysics with the experimental method, discuss the physical theories by reasons drawn from metaphysical systems, include in your metaphysical systems theories of the positive sciences. The positivist will have no difficulty in proving to you that the physical methods cannot lead to consequences that you wished to draw from them and will conclude that the principles of metaphysics cancel one another. He will have no trouble in showing that your metaphysical deductions cannot compete with physical theories resting on experimental laws, and from there he will conclude that metaphysics is condemned by its very consequences.

If you do not establish a radical separation between physics and metaphysics, if you confuse them, you will be forced to recognize that the method of physics is good even in metaphysics. But this is to give the advantage to positivism.

6

The Primacy of Common Sense.

From "Quelques réflexions au sujet de la physique expérimentale," *Revue des questions scientifiques* 36 (1894), pp. 211-14 **(1894-5)**.

As an experiment in physics is something very different from a simple registering of a fact, one can easily see that the truth and certainty of an experimental result are of a very different order from the truth and certainty of a fact being registered and that these certainties, so different, about nature are appraised by methods altogether distinct.

When a sincere witness, sufficiently sober so as not to take the whims of his imagination for observation and familiar with the language he uses to express his thought clearly, affirms to have registered a fact, the fact is certain. If I declare to you that on such a day, at such an hour, in such a street of the city I saw a white horse, you must believe, unless you have reason to consider me a liar or a victim of hallucination that on that day, at that hour, in that street, there was a white horse.

The confidence that should be accorded to the proposition stated by a physicist as the result of an experiment is not of the same nature. If the physicist had limited himself to listing the facts which he has seen, and with his very eyes, and seen it in the strict sense of that word, his witness would have to be evaluated by the general rules appropriate for establishing the degree of confidence deserved by a man's witness. If the physicist was known to be trustworthy — and this, I think, would be the general case — his testimony would have to be received as an expression of truth.

But once more, what the physicist states as the result of an experiment is not a recital of facts registered by him. It is rather the interpretation of these facts, it is their transposition into the abstract, symbolic world created by theories which he considers to be well established.

Therefore, after the physicist's testimony has been submitted to the rules which establish the degree of confidence due to the account of a witness, you have accomplished only a part, and the easier part, of the critique of his experiment.

It is necessary, in the first place, that you inquire with great care about the theories which he regards as established and which he used in interpreting the facts registered by him. Unless you know these theories, it will be impossible for you to grasp the meaning which he attributes to his statements. This physicist would in that case be to you as a witness to a judge who does not understand his language.

If the theories accepted by that physicist are the same you are accepting, if you agree to follow the same rules in interpreting the phenomena, then you speak the same language and you can make yourself understood. But it is not always so. It is not so when you discuss the experiments of a physicist whom half a century, a century, or two centuries separate from you. It then becomes necessary to establish a correspondence between the theoretical ideas of the author you are studying and your ideas, to interpret anew, by means of symbols you are accepting, what he had interpreted by means of symbols accepted by him. If you succeed, it will be possible for you to discuss his experiment. This experiment will be a testimony rendered in a language foreign to you, but in a language whose dictionary you possess. You can examine it.

Newton, for example, has made certain observations touching on the phenomenon of colored rings. He had interpreted these observations in terms of the optical theory created by him, that is, according to the theory of emission. He interpreted them as giving, through the luminous corpuscles of each color, the distance between the point of an easy reflexion and of an easy transmission. When Young and Fresnel reintroduced the wave theory in order to replace with it the theory of emission, it was possible for them to make parts of the new theory correspond in certain respects to parts of the old theory. They saw, in particular, that the distance between the point of an easy reflexion and the point of an easy transmission corresponded to a fourth of what the new theory called a wavelength. Thanks to that remark, the results of Newton's experiments could be translated into the language of undulations. The numbers given by Newton, when multiplied by four, gave the wavelengths of different colors.

If, on the contrary, you cannot obtain sufficient information about the theoretical ideas of the physicist whose experiments you are discussing, if you do not succeed in establishing a correspondence betwen the symbols he had adopted and the symbols that carry the theory accepted by you, if you cannot translate into your language the propositions by which he had represented the results of his experiments, these results will be neither true, nor false. They will be void of meaning. They will be dead letter. How many observations, accumulated by physicists of old, have thus fallen into oblivion! Their authors have neglected to inform us about the methods whereby they interpreted the facts. It is impossible for us to transpose their intepretations into our theories. They have enclosed their ideas within signs to which we do not have the keys.

The first rules may seem to be naive and one may wonder on seeing us insist on them. However, if these rules are banal it is even more banal to overlook them. Great is the number of scientific discussions where each of the two protagonists tries to vanquish his opponent by the irrefragable witness of facts. Actually, contradictory observations are opposed to one another. The contradiction does not exist in the facts as they are always in accord with one another. The contradiction is between the theories through which each of the two protagonists express the reality. Great is the number of propositions taken for monstrous errors in the writings of those who have preceded us. One would perhaps celebrate them as great verities if one were willing to become well informed about the theories that would give those propositions their true meaning, if one cared to translate them into the language of theories favored today.

But let us suppose that you have registered the agreement between the theories admitted by an experimenter and the theories you consider as exact. You are far from being in the position to make your own the judgments by which he states the results of his experiments. You must now examine whether in the interpretation of the observed facts he had correctly applied the rules given by the theories accepted by both of you, whether he had made all the necessary corrections. All too often you will find that the experimenter has not satisfied all the legitimate demands. In applying his theories he has committed an error of reasoning or of calculation, or he has omitted an indispensable correction and let an error remain present that could have been eliminated.

The experimenter has used, in order to interpret his observations, theories which you accept as well as he does. He has correctly applied, in that interpretation, the rules prescribed by those theories. He has eliminated the causes of errors or has corrected their effects. Still this is not enough to adopt the results of his experiments. The abstract propositions which the theories make correspond to the observed facts are not, we have said, entirely specified. To the same facts there may correspond an infinity of different propositions, an infinity of evaluations that are expressed by different numbers. The degree of possible indetermination of the abstract mathematical proposition by which the results of an experiment are conveyed, is the degree of approximation of that experiment. You must know the degree of approximation of the experiment which you examine. If the experimenter has indicated it, you must first be assured of the reasonings that helped him to evaluate it. If he did not indicate it, you have to determine it by your own discussions. This appraisal of the degree of observation implied in a given experiment is a delicate operation. Very often it is so complicated that a wholly logical order is difficult to maintain. Reasoning must then yield to that rare subtle quality, to that sort of touch which is called experimental sense —

a property more of the intuitive mind than of the geometrical mind.

The simple description of rules that preside over the examination of an experiment in physics, over its acceptance or its rejection, is sufficient to bring out this essential truth: The result of an experiment in physics does not possess a certitude of the same order as does a fact registered by non-scientific method, that is, by the plain look or plain touch of a man sane in body and mind. Being less immediate and subject to discussions that cannot touch on commonsense testimony, the certitude [of an experiment] remains always subordinate to the confidence inspired by a whole ensemble of theories.

7

Catholic Patriotism

From *Le Nouvelliste* (Bordeaux), 28 juin, 1899, p. 3. cols. 4–5 (**1899–15**). Translation first published in *Uneasy Genius*, pp. 155–56.

My dear Camarades,

You will allow me to address you with this word because, even though we did not sit on the same benches, we have been formed by the same teachers and it is to this common education, a firm assurance of common sentiments, that I owe the honor of presiding over your festive reunion.

Indeed, the influence of the great moulders of men who have distinguished the Society of Mary made its impact on all of us; although in different places, that influence has implanted in all of us the same principles capable of maintaining forever the harmony of our thoughts and the unity of our efforts. At this hour when the winds from the four corners of the world blow divisiveness and hatred, it is appropriate to point out clearly the ties which unite us into a closely knit alliance.

We admire, we love, we serve the same things and these things are the ones which are symbolized by the blazon of our Collège Stanislas.

You are familiar with that blazon: one half of it is occupied by a book, the other half by a knight armed from top to toe; joining the two is the emblem of France.

The book stands for all the truth, beauty, and goodness produced by the thought of all peoples and centuries, but especial-

ly for the products of the minds of the Greeks and Romans —
educators of our national genius and especially of the French way
of thinking, which is the clearest, most precise and logical, and, at
the same time, the most humane way of thinking in the modern
world. This is what our teachers have in the first place taught us
to savor. Their efforts have not been sterile. From the moment
they gave a Sainte-Claire Deville to the world of science to the
day they explained *Le Cid* to the future author of *Cyrano de Berge-
rac*, they have launched a good number among us into those
peaceful conquests of the intellectual world which enlarged France
by making larger the possessions of mankind.

Beside the book there is the knight well set on his charger,
his sword drawn. What was seen in him by the ardor of our
eighteen-year-olds, was not the brain of a France with clear ideas
but the strongly beating heart and the boiling blood of France, the
Army.

The Army! I cannot pronounce this word without seeing
again at my side absorbed in silent study camarades and friends
who prepared by severe efforts for the honor of carrying the
sword: a Berger, who in Tonkin would unite his blood with that
of Négrier; a Duchâtelet, who would enter Tananarive on the side
of Duchêne; a De Planhol, who exhausted by fever would collapse
on the burning sand to die there as died the valiant knights of
Tunis in the time of Saint Louis.

When these, on their first leave from St. Cyr, walked across
the courtyards of the Collège Stanislas, proud of the brand new
cassowary plumes which clapped on their shakos, there were in
those courtyards others, their juniors, who vowed to imitate and
outdo them. Among them were Gourand, who was to lay firm
hold on the unconquerable Samary, and Barattier, loyal aide of
Marchand in that expedition which could make jealous the ten
thousands of Xenophon if jealousy were not the lot of the living.

Between the book and the knight is the emblem of France as
if to vivify by the same breath, as if to fuse in one single idea and
in the same love every field of science, every beauty of literature,
all bravery of the Army. How well placed were, in the center of
our blazon, those three white lilies against an azure field, symbol

of an education in which everything tended to make us know and love France in each of her soil's regions, the France of Clovis and of Charlemagne as well as the France of Saint Louis and Joan of Arc, the France of Henri IV and of Louis XIV as well as the France of Napoleon, the France victorious at Valmy and Iena as well as the France murderous at Saint-Privat and at Patay.

To know and to love one's country is something but not all. It is necessary to serve it and to contribute effectively to its prosperity and glory. Our teachers knew this and sought to make us men capable of accomplishing this task without fail.

They wanted us, first of all, to be men of initiative. To have initiative is not merely to propose a target for one's activity. Initiative consists above all in keeping one's will firm in the teeth of adversities, temptations, and discouragements along the way one has chosen. Initiative consists in obeying for a whole lifetime the order which one has imposed on oneself. Therefore, in order to learn how to use our will, our teachers taught us to obey; they bent us along the lines of discipline — without which the will becomes caprice — strict, exact, specific discipline, but a discipline accepted loyally and gladly, because it was right, steady, free of surprises and sudden moves, especially because those who imposed it upon us have assumed a stricter yoke and preached by example.

In the life of a man of initiative there are grave hours, hours when he has to choose between happiness and mission, hours when he has to sacrifice himself. Our teachers foresaw those hours and inspired in us the spirit of sacrifice. Spirit of sacrifice! What a sense these words assumed when they fell from the lips of the Abbé de Lagarde as he let lightning flash from under his eyelids, already half closed for the grave, when he stiffened for a moment his body bent much too soon and ravaged by the unspeakable pains of cancer!

Cult of science and letters, patriotism, spirit of discipline and of initiative, spirit of sacrifice — to let these sentiments germinate and grow in us our teachers relied on the help of the One who reinforced the heart of man. In every truth, and beauty, they showed us the reflection of eternal Truth and of a supreme Beauty. In the annals of the history of France — of her intellectual history as well as in her military history — they taught us to perceive the gesture — conscious and unconscious — of God's soldiers. To bend our excesses under the yoke of discipline, they taught us that all authority comes from God; to kindle in us the spirit of sacrifice, they constantly set before us the image of the Crucified God. To give France "Frenchmen without fear" they exerted themselves to give the Church "Christians without blame."

My dear Camarades, I hope to have faithfully traced the characteristic features of the education we have received. If we have come together today, it is to assert that we are proud of having received that education, to acknowledge that the experience of life has made us understand its price increasingly every day, and to thank those to whom we are indebted for it. It is on this note of gratitude that I want to end. I raise my glass in honor of our teachers, priests and brothers of the Société de Marie, and in order to make our homage more concrete, I propose two toasts: one in honor of the one whom I saw in the days of my childhood direct my seniors at Stanislas and whom I see today presiding with so much vitality over the frolics of the children of Sainte-Marie, to our dear Mr. Hérail; — the other to the one who directs so actively Sainte-Marie, our Stanislas of the Southwest, to my former fellow student, the Abbé Bernard.

8

Evolutionary Competition of Ideas

Review of *L'Avenir du Christianisme. Introduction: La vie et la Pensée chrétienne dans le passé,* par Albert Dufourcq, professeur à l'Université de Bordeaux (Paris: Bloud et Cie, 1904), x + 779pp. *Revue des questions scientifiques* 55 (1904), pp. 252–60 **(1904–24)**.

I. Pascal wrote: "It is in vain to see, with the eyes of faith, Darius and Cyrus, Alexander the Great, the Romans, Pompey, and Herod act, without knowing that they acted for the glory of the Church."

This *pensée,* as Sainte-Beuve recalled, contained in seed [Bossuet's] *Discours sur l'Histoire universelle,* a robust tree which issued from that small grain.

The idea which that *pensée* contains has for a long time haunted the minds of [non-]Christian historians. The entire evolution of mankind appeared to them as an undecipherable enigma to which one does not have the key. And the key is Jesus. The preparation for the coming of the Son of God, the preaching of his Gospel, the development of his Church, these are the filament that guides the reason in the maze of the facts of history and reveals a plan of admirable unity in what could seem but a work of chance.

The work of chance, or rather the inextricable texture of fatal consequences produced by the entanglement of laws with no purpose, this is what rationalism sees in the history of humanity.

The evolution of an animal species endowed with a brain of complicated convolutions which an unconscious selection rendered more apt than other primates to react against the causes of destruction surrounding him, such as heat, cold, hunger, the ferocity of carnivores; the evolution of a species whose different families kill one another so that natural selection may achieve its progressive work and allow to survive only the select races; the evolution of a species where a few individuals, inferior in muscles but superior by cerebral development, had recourse to ruse to tame their robust adversaries and have enchained their instincts by means of fallacies called *morality* and *religion; the evolution of a species*, where others, in order to master more effectively the forces of nature, thought up the sciences whose sole legitimate object is the increase of the doses of physical enjoyments to be shared by each representative of the species — enjoyments which the strongest and the most skillful monopolize while the miserable crowd labors and suffers in the expectation of a violent revolution which will allow them to satiate themselves; the evolution of a species with no goal for the individual whom the chemical actions dissolve after a few years during which he knew more bitterness than joy; the evolution with no goal for the species whose last representatives will die of cold and hunger on a frozen planet where no geologist will exhume their remains. Such is history for the one who wants to approach it as a zoologist and not as a Christian.

This is not the history which Albert Dufourcq sets forth for us. With Cournot, whose profound remark he recalls, he admits that "the religion which our fathers have handed down to us is not like another religion. That religion fulfills in the history of the civilized world a unique role, a role with no equivalent or analogy." This is the role which he undertakes to describe to us, since the times lost in the mist of prehistory until the dawn of revolutionary torments. More precisely, he aims at showing that "the end of history is the realization of a common conscience of humanity and that Christianity is the form of that universal conscience."

It is not as points of a philosophical dissertation that this great idea unfolds under our very eyes. According to the method dear to our modern Historical School this great idea does not try to formulate itself in abstract and general propositions. It shows itself to us as it has developed in the world, concretely and alive. It will speak through the mouth of those who had for their mission to teach humanity. It will vibrate in the tremors of popular pressures, commotions, and revolutions. One will hear it run under the thick layers of events.

The words of men or the recital of facts, all this has been subjected to the crucible of severe criticism. Trained in rigorous discussion of history, disciple of a School where one is not easily led either by traditions of nebulous origin or by apocryphal documents, Mr. Dufourcq, in order to trace out the triumphal march of the Christian idea, accepts only markers of granite on which the teeth of scepticism are readily broken. And he marshals such a multitude of authentic facts that press upon one another, all oriented toward the same goal that it would be childish to tarry and discuss a doubtful event or a suspect testimony.

II. The description of the pre-messianic times is but an introduction in Mr. Dufourcq's book. It corresponds to the opposition between Judaism and paganism. Judaism is a faithful sentinel who watches over the safeguarding of two ideas: the divine paternity and the human brotherhood. Paganism is the scoffer at those two ideas by turning the divine Father into an inexorable destiny, into a monstruous tyrant or into a participant of debaucheries and transforming thereby brother-man into a slave.

In the midst of a deaf anxiety which turns Judaism and paganism toward the coming of new ideas, the Promised One appears. Humble workman of a small country, He goes by, curing the blind and the lame, raising the dead, and evangelizing the poor. He dies ignominiously, victim of all sorts of jealousy and hypocrisy. But the death on the cross is only the prelude of the triumph of the Resurrection and Ascension.

And immediately follows the development of Christ's Church, the perpetual renewal of the miraculous drama that crowned the life of its Head on earth, a continual sequence of passions that lead to death only to terminate in resurrection.

The small Christian community of Jerusalem is destroyed by persecution. But the wine-press that crushes it splashes far and wide the Christian idea whose scattered droplets will make paganism ferment. Churches are established in Antioch, in the various towns of Asia Minor, in Greece, in Rome. Paul of Tarsus seizes Christ's work from the avaricious narrowness of Judaism and proclaims the catholicity of Christianity. Peter installs the seat of the nascent papacy in the capital of Caesars. John, the centenarian apostle, bars the inroads of heresies that advance by composing a mystical Gospel. Irenaeus, the father of apologists, clarifies the Christian doctrine and crowns thereby the messianic times.

III. From the 3rd to the 11th century extends what Mr Dufourcq calls the *mediterranean epoch*. "The mediterranean lands, and they only, see the development of the life Christ brought," the Apostles preached, and the successors of Irenaeus organized.

A first period, that of Roman Christianity, covers the 3rd and 4th centuries.

Two powers confront one another: the Christian Church and the pagan Empire. Their principles oppose one another as do negation and affirmation. Between the two no reconciliation is possible: "*This* will kill *that*," and with Septimus Severus the Roman Empire certainly thinks to be *this* which will kill *that*, the Church of Christ. The persecution is unleashed with alternating periods of relative calm and of unheard cruelties. The faith of martyrs and the charity of Christians get the better of the Empire whose defeat is accepted by Constantine and who can cry out with the vanquished Julian: "You have won, Galilean!"

"The alliance of two powers, former enemies, establishes between them close rapport which gives rise to struggles of influence. The imperial traditions are first the stronger, afterwards the Christian traditions prevail" by purifying the manners of pagan corruption and by mitigating with Christian charity the harsh legislations of Antiquity. The Church, now free of interference,

develops its own organization. The virtues of ascetics sweeten
these times that see the birth of the first monasteries.

While the struggle between the Church and the Empire
unfolds, a struggle which history will see arise again and again,
there emerges another struggle which also will witness endlesss
recurrences. The revealed doctrine clashes with an independent
science. At that moment the dispute is particularly intense between
Hellenistic science that radiates around the School of Alexandria
and the theology of the apologists. As autonomous reason feels
assured of victory over revealed dogma, two currents form. In
various epochs of history we shall see them advance for the great
advantage of secular science as well as of sacred doctrine. One of
those currents — *intellectus quaerens fidem* — makes Greek science
Christian; the other — *fides quaerens intellectum* — will organize
Christian theology according to the demands of Greek dialectic.
These two currents join in a synthesis to which Saint Augustine's
great name is attached.

But a new passion comes and terrible trials hit the Church.

Constantinople, which took the place of Rome as the capital
of the Empire, tries to gain the Christian primacy as well. The
separatist mentality which develops there prepares the schism. The
vivifying strength that enlarges the trunk ceases to breathe on the
cut-off branch. The latter becomes enervated in dissolute manners,
childish ambitions, quarrels with no objectives until the day when
the Turks set fire to that branch, dry for a long time.

On the other hand, Islam stirs up the Arab tribes as the desert
wind whips up the sand. It unleashes them in a furious torment
across Africa, Spain, Aquitaine. While Greek science is reduced to
ashes in Alexandria, the glorious churches of Cyprian and Augustine are swept away. The Christian idea disappears from the
regions that saw its birth, from the very regions where it once
most strongly resonated.

Is the Christian idea dead? Not at all! Driven from the East
and from the South by the Muslim hurricane, it advances in wide
streams toward the north and the west. As a tide that rises with no
letup, it overflows the newly arrived barbarians of Germania,
watering their paganism or Arianism. It overflows Ireland which

St Columba and his monks conquer forever for the Church. Its last waves will produce saints in the lands of Swabia and Westphalia, impregnates with the life of the Gospel the distant regions inhabited by the Slavs and the Hungarians. But its deepest waters cover the land of the Franks. To the *period of westward march* succeds the *Frank period dominated by Charlemagne's colossal stature.*

Momentary triumph of the evangelical idea, soon followed by a new period of trial.

The invasions of new barbarians, the rekindling of the instincts, not yet under complete control, of the newcomers, break the Frankish empire. "Once the emperor is gone, the kings disappear one by one. Pity the one who did not know how to gather behind him a strong association of loyal vassals. . . . Pride fells the strong, fear brings down the week, egotism corrodes all souls. . . . To live from day to day, as one can manage, each one for himself; to escape massacres and famines, such is the whole aim of these poor humans degraded by the fear of sickness, conflagration, and death." They see with terror the coming of the year 1000, the last one for the world.

This anarchy puts religious life at a grave peril. The lords confiscate the goods of the Church or abuse its dignitaries. Simoniac clergy live in disorder. Is this the end of the Christian idea? In that profound darkness it casts but a pale and trembling gleam. Yet it continues, in many places, to shine in the hearts of numerous saints. Threatened by extortions of all sorts, the weak look for a sovereign strong enough to protect them, sufficiently loyal not to oppress them, and not finding one on earth, they bequeath their goods to the Apostle. They thus create the patrimony of the Church of Rome whose formation and evolution have been studied by my lamented good friend Paul Fabre. Finally the monastery of Cluny is founded, a foyer where the little Christian flame revives and becomes a burning torch.

IV. Here, at the end of the 11th century, when the Orient definitively separates from the Church of Rome, the latter resumes its powerful and salutary influence on the western nations of Europe.

The energetic action of the papacy, now in the hands of such men as Gregory VII, the fervent preaching of monks like Saint Bernard, restore to the Church the purity of mores and the exactness of discipline. Once again healthy and strong, the Church conquers the feudal society which it supervises and organises. To push back the Muslims the Church makes the most of the excessive violence and eagerness to fight which boils in that society.

Before this power of the Christian idea which the papacy personifies, the civil powers, all resentful of their humiliation, are forced to bow. Emperor Henry IV does penance at Canossa and Barbarossa's pride is broken by Alexander III. While Saint Francis and Saint Dominic rekindle the faith in the souls and revive the virtues of the first Christians, Christian thought resumes the work of Saint Augustine. Saint Anselm, Hugo of Saint Victor, and Petrus Lombardus initiate a theological and philosophical tradition which produces its masterpiece in the work of Saint Thomas Aquinas. This is the moment when the secular sciences resume their progress with Albert the Great, Roger Bacon, Petrus Peregrinus, Jordanus Nemorarius, when gothic art raises its purest cathedrals, when the king of France is Louis IX. With that glorious *period of feudal Christendom* opens the *Western epoch*.

But it is here that opens up a *period of warfare for the Church*. The transfer of the papacy to Avignon prepares the schism of the West. The struggle of the princes against papal power resumes with a new intensity as their administrative work attacks the *de facto* independence of the Church. D'Ailly and Gerson, Wycliff and John Huss revive the heresies. Their statist or individualist theories attack the speculative idea of the Church. During these times profound upheavals uproot ideas which one believed had rested on unshakable foundations. Columbus, Vasco de Gama, and Magellan transform and enlarge the notion man had formed about habitable lands. Copernicus teaches him that his abode is but a planet, similar to the others, that it is neither the center of the world, nor the axis of the celestial motion. The nascent printing press spreads with abundance these novel opinions while at the same it exhumes the pagan thought that had been buried for centuries. The papacy tries to impregnate with Christian tendency the Renaissance art, but the Renaissance paganizes the Church by weakening and corrupting it. A Borgia sits on the chair of Saint Peter.

On October 31, 1517, Luther opens the quarrel of indulgences. In 1527 the soldiers of the Bourbons set upon Rome with fire and sword.

The Reformation spreads across Europe. It seizes vast territories from the supremacy of the Church of Rome. In many places the princes ally themselves with the religious revolt and enlarge their civil authority from the debris of spiritual authority which they had coveted for a long time.

But the Church recognizes the magnitude of the problem and makes ready to remedy it. Paul III chooses a commission charged with the preparation of reform and hands to each of its members the following instruction: "We have chosen you in the hope that you may reestablish in our hearts and in our works the authority of Christ already forgotten by the lay people and by us, members of the clergy; that you may bring remedies to our maladies; that you may lead the sheep of Christ back to their only fold and that you may remove from us God's ire and vengeance. We have deserved it and it is on the point of falling upon us."

"Supported by the defensive militia of the Inquisitors and by the offensive militia of the Jesuits, the papacy," so Professor Dufourcq writes, "imposes the reform on the Catholic world. . . The papacy organizes, directs, and protects the ecumenical Council of Trent which solidifies the Church on its basis." Against the anarchical multiplicity of Protestant beliefs the disciplined unity of Catholic dogma reasserts itself. The advances of Protestantism are stopped and in countries that remained united, faith and charity enlighten and inspire the great 17th century.

However, new dangers are not late in coming. Not only does Protestantism stand firm on its positions from which the Church does not succeed in dislodging it, but in Catholic countries the two perennial enemies once more confront one another. Elated by the dazzling progress of the positive sciences from Galileo to Descartes and from Huygens to Newton, the rationalist disbelief deprives the Church of its intellectual elite. Moreover, the State once more tries to absorb the Church. Monarchs of Catholic countries, incited by the example of Protestant princes, want to dominate the spiritual power. Among the dangers that threaten the Christian idea within Catholic monarchies this is perhaps the gravest. The French Revolution will break that spiritual power.

V. After having drawn in broad strokes and in vivid and vigorous colors the picture of which we have tried to reproduce a pale sketch, Mr Dufourcq distances himself from his work in order to embrace it in a single glance and receive its general view.

Here are the terms in which he does this: "In the past, the Christian development did not appear similar to the development of a logical system which, once its principle is posed, would draw its consequences by the deductive approach."

"The Christian development does not appear something analogous to the regular evolution of a living germ which grows, expands, and dies. . . . The Christian development presents an original and unique physiognomy. It is not subject to laws—nor is it left to chance. One cannot classify it either as arbitrary or necessary. It appears as something conditioned by an *infinity* of facts which, in last analysis, derive from man's free will and from the free grace of God."

9

Physics of a Believer

Quoted from P. Duhem, *The Aim and Structure of Physical Theory*, tr. Philip
P. Wiener (Princeton, N.J.: Princeton University Press, 1954), pp. 273–79,
287–90, 305–11. Reprinted with permission. For the French original, see
1905–8 and **1914–2**.

1. INTRODUCTION. The *Revue de métaphysique et de morale* a
little more than a year ago   published an article in which the
opinions I emitted on different occasions concerning physical
theories were expounded and discussed. The author of that article,
Abel Rey, has taken the trouble of studying assiduously even the
smallest writings in which I had expounded my thought, and he
has followed the course of this thought with a great concern for
accuracy; thus, he has drawn for his readers a picture whose
fidelity has keenly impressed me; and surely, I shall not be
bargaining with Mr Rey by offering him acknowledgment of my
appreciation in exchange for the sympathy with which his
understanding has assimilated what I had published.

And yet (is there anyone who does not find something about
which to complain in his own portrait, however accurate the
painter may have been?), it seemed to me that Mr Rey has
solicited somewhat more than exactly the premises which I had set
down and that he has drawn conclusions from them which were
not altogether contained in them. I should like to apply some
restrictions to these conclusions.

Mr Rey terminates his article as follows:

"Our intention here has been to examine only the scientific philosophy of Mr Duhem and not his scientific work itself. In order to find and formulate precisely the expression of this philosophy . . . , it seems that we may propose the following formula: In its tendencies toward a qualitative conception of the material universe, in its challenging distrust with regard to a complete explanation of this universe itself, of the sort of mechanism he imagines it has, and in its animadversions, more pronounced than genuine, with respect to an integral scientific scepticism, Duhem's scientific philosophy is that of a believer."

Of course, I believe with all my soul in the truths that God has revelead to us and that He has taught us through His Church; I have never concealed my faith, and that He in whom I hold it will keep me from ever being ashamed of it, I hope from the bottom of my heart: in this sense, it is permissible to say that the physics I profess is the physics of a believer. But surely it is not in this sense that Mr. Rey meant the formula by which he characterized this physics; rather did he mean that the beliefs of the Christian had more or less conciously guided the criticism of the physicist, that they had inclined his reason to certain conclusions, and that these conclusions were hence to appear suspect to minds concerned with scientific rigor but alien to the spiritualist philosophy of Catholic dogma; in short, that one must be a believer, not to mention being a perspicacious one, in order to adopt altogether the principles as well as the consequences of the doctrine that I have tried to formulate concerning physical theories.

If that were the case, I should have been singularly pursuing the wrong course and failed of my aim. In fact, I have constantly aimed to prove that physics proceeds by an autonomous method absolutely independent of any metaphysical opinion; I have carefully analyzed this method in order to exhibit through this analysis the proper character and exact scope of the theories which summarize and classify its discoveries; I have denied that these theories have any ability to penetrate beyond the teachings of experiment or any capacity to surmise realities hidden under data observable by the senses; I have thereby denied these theories the power to draw the plan of any metaphysical system, as I have denied metaphysical doctrines the right to testify for or against any physical theory. If all these efforts have terminated only in a conception of physics in which religious faith is implicitly and almost clandestinely postulated, then I must confess I have been strangely mistaken about the result to which my work was tending.

Before admitting such a mistake, I should like to be allowed to glance again at this work as a whole, to fix my gaze particularly on the parts in which the seal of the Christian faith was believed noticeable, and to recognize whether, against my intention, this seal is really impressed therein or else, on the contrary, whether an illusion, easy to dissipate, has not led to the taking of certain characteristics not belonging to the work as the mark of a believer. I hope that this inquiry, by clearing up confusions and ambiguities, will put the following conclusions beyond doubt: Whatever I have said of the method by which physics proceeds, or of the nature and scope that we must attribute to the theories it constructs, does not in any way prejudice either the metaphysical doctrines or the religious beliefs of anyone who accepts my words. The believer and the nonbeliever may both work in common accord for the progress of physical science such as I have tried to define it.

. . . .

2. OUR PHYSICAL SYSTEM IS POSITIVIST IN ITS ORIGINS

We should like to prove that the system of physics which we propose is subjected in all its parts to the most rigorous requirements of positive method, and that it is positivist in its conclusions as well as in its origins.

First, of what preoccupations is the constitution of our system the result? Is our conception of physical theory the work of a believer who is uneasy about the disparity between the teachings of his church and the lessons of reason? Does it arise from an effort that faith in divine things would have attempted in order to attach itself to the doctrines of human science (*fides quaerens intellectum*)? If so, the nonbeliever may conceive legitimate suspicions regarding such a system; he may fear that some proposi-

tion oriented toward Catholic beliefs has, unawares even to the
author, slipped through the close meshes of rigorous criticism, so
ready is the human mind to think true what it wishes! On the
other hand, these suspicions would cease to have any ground if the
scientific system occupying us were born within the very matrix
of experiment and were forced on the author outside of any
metaphysical or theological concern, and almost despite himself,
through the daily practice and teaching of science.

Here then we are going to relate how we were led to teach
concerning the aim and structure of physical theory an opinion
that is said to be brand new; we shall do so in all sincerity, not
because we have the vanity to believe the career of our thought
interesting in itself, but in order that the knowledge of the origins
of the doctrine may make for a more exact judgment of its logical
validity, for it is this validity that is in question.

Let us take ourselves back twenty-five years to the time when
we received our first initiation, as physicist-to-be, in the mathematics classes of the Collège Stanislas. The man who gave us this
initation, Jules Moutier, was an ingenious theorist; his critical
sense, ever alert and extremely perspicacious, distinguished with
sure accuracy the weak point of many a system which others
accepted without dispute; proof of his inquiring mind is not
lacking, and physical chemistry owes one of its most important
laws to him. It was this teacher who planted in us the seed of our
admiration for physical theory and the desire to contribute to its
progress. Naturally he oriented our first tendencies in the same
direction to which his own preferences brought him. Now,
although he appealed in his investigations to the most diverse
methods, each in turn, it was to the mechanical attempts at
explanation that Moutier returned most often with a sort of
predilection. Like most of the theorists of his time he saw the ideal
of physics in an explanation of the material universe constructed
in the manner of the atomists and the Cartesians; in one of his
writings he did not hesitate to adopt the following thought of
Huygens: "The causes of all natural phenomena are conceived
through mechanical reasons, unless we wish to give up all hope of
understanding anything in physics."

Being a disciple of Moutier, it was as a convinced partisan of mechanism that we approached the course in physics pursued at the Ecole Normale. There we were to come under influences very different from those we had experienced until then; the jesting scepticism of Bertin struck in vain against the constantly reborn and constantly abortive attempts of the mechanists. Without going as far as to have the agnosticism and empiricism of Bertin, most of our teachers shared his mistrust regarding hypotheses about the intimate nature of matter. Past masters in experimental manipulation, they saw in experiment the only source of truth; when they accepted physical theory it was on condition that it rest entirely on laws drawn from observation.

Whereas the physicists and chemists rivalled one another in praising the method that Newton had formulated at the end of his book of *Principia*, those who taught us mathematics, especially Jules Tannery, worked to develop and sharpen in us a critical sense and make our reason infinitely difficult to satisfy when it had to judge the rigor of a demonstration.

The tendencies which the instruction of the experimenters had produced in our mind and the lessons that the mathematicians had fixed in us concurred in making us conceive physical theory to be of quite a different type from what we had imagined it to be until then. This ideal theory, the supreme goal of our efforts, we wished to see resting solidly on laws verified by experiment and completely exempt from those hypotheses about the structure of matter which Newton had condemned in his immortal *General Scholium*; but at the same time we wished theory to be constructed with that logical rigor which the algebraists had taught us to admire. It was to the model of such a theory that we tried hard to make our lessons conform when we were given the first opportunity to teach.

We soon had to recognize how vain our efforts were. We had the good fortune to teach before an elite audience in the Faculty of Science at Lille. Among our students, many of whom are today colleagues of ours, the critical sense was hardly asleep; requests for clarification and embarrassing objections indefatigably indicated to us the paradoxes and vicious circles which kept reappearing in our lessons despite our care. This harsh but salutary test did not take long to convince us that physics could not be constructed on the plan we had undertaken to follow, that the inductive method as defined by Newton could not be practised, that the proper nature and true object of physical theory had not yet been exhibited with complete clarity, and that no physical doctrine could be expounded in a fully satisfactory manner so long as this nature and object had not been determined in an exact and critical manner.

This necessity to take up again the analysis of the method by which physical theory can be developed, down to its very foundations, appeared to us in circumstances of which we retain a very vivid recollection. Little satisfied with the exposition of the principles of thermodynamics that they had encountered "in books and among men," several of our students asked us to edit for them a small treatise on the foundations of that science. While we tried hard to satisfy their desire, the radical impotence of the methods then known for constructing a logical theory came home to us more persistently each day. We then had an intuition of the truths which since that time we have continually affirmed; we understood that physical theory is neither a metaphysical explanation nor a set of general laws whose truth is established by experiment and induction; that it is an artificial construction manufactured with the aid of mathematical magnitudes; that the relation of these magnitudes to the abstract notions emergent from experiment is simply that relation which signs have to the things signified; that this theory constitutes a kind of synoptic painting or schematic sketch suited to summarize and classify the laws of observation; that it may be developed with the same rigor as an algebraic doctrine, for in imitation of the latter it is constructed wholly with the aid of combinations of magnitudes that we have ourselves arranged in our own manner. But we also understood that the requirements of mathematical rigor are no longer relevant when it comes to comparing a theoretical construction with the experimental laws which it claims to represent, and to judging the resemblance between the picture and the object, for this comparison and judgment do not arise from the faculty by which we can unwind a series of clear and rigorous syllogisms. We realized that in order to judge this resemblance between theory and empirical data, it is *not* possible to dissociate the theoretical construction and to submit each of its parts in isolation to the test of facts, for the slightest experimental verification puts into play the most diverse chapters of theory, and we realized that any comparison between theoretical physics and experimental physics consists in an alliance of theory taken in its entirety with the total teaching of experiment.

It was thus through the necessities of teaching, under their urgent and constant pressure, that we were led to produce a conception of physical theory markedly different from what had been current till then. These same necessities led us through the years to develop our first thoughts, to make them more precise, to explain and to correct them. It was through these necessities that our system concerning the nature of physical theory was affirmed in our conviction, thanks to the ease with which it enabled us to connect into a coherent exposition the most diverse chapters of science. And may we be pardoned for insisting here on indicating the quite special authority conferred on our principles by this test to which we have submitted them in the course of many long years? There are many persons today who write about the principles of mechanics and physics, but if someone proposed to them that they give a complete course in physics which would still agree in all particulars with their doctrine, how many of them would accept the challenge?

Our ideas about the nature of physical theory are, therefore, rooted in the practice of scientific research and in the exigencies of teaching. Deeply as we have gone into our examination of our intellectual conscience, it is impossible for us to recognize an influence exerted on the genesis of these ideas by any religious preoccupation whatever. And how could it be otherwise? How could we have imagined that our Catholic faith was interested in the evolution undergone by our opinions as a physicist? Have we not known Christians, as sincere as they were enlightened, who firmly believed in the mechanical explanation of the material universe? Have we not known some of them to be ardent partisans of the inductive method of Newton? Was it not a glaring fact to us, as to any man of good sense, that the object and nature of physical theory are things foreign to religious doctrines and without any contact with them? And, furthermore, as though better to mark to what little extent our manner of viewing these questions was inspired by our religious beliefs, have not the most numerous and liveliest attacks against this manner of viewing things come from those who profess the same faith as we do?

Our interpretation of physical theory is, therefore, essentially positivist in its origins. Nothing in the circumstances which suggested this interpretation can justify the distrust of anyone who does not share our metaphysical convictions or religious beliefs.

. . . .

5.   OUR SYSTEM DENIES TO PHYSICAL THEORY ANY METAPHYSICAL OR APOLOGETIC IMPORT

That our physics is the physics of a believer is said to follow from the fact that it so radically denies any validity to the objections obtained from physical theory to spiritualistic metaphysics and the Catholic faith! But it might just as well be called the physics of a nonbeliever, for it does not render better or stricter justice to the arguments in favor of metaphysics or dogma that some have tried to deduce from physical theory. It is just as absurd to claim that a principle of theoretical physics contradicts a proposition formulated by spiritualistic philosophy or by Catholic doctrine as it is to claim that it confirms such a proposition. There cannot be disagreement or agreement between a proposition touching on an objective reality and another proposition which has no objective import. Every time people cite a principle of theoretical physics in support of a metaphysical doctrine or a religious dogma, they commit a mistake, for they attribute to this principle a meaning not its own, an import not belonging to it.

Let us again explain what we are saying by an illustration.

In the middle of the last century, Clausius, after profoundly transforming Carnot's principle, drew from it the following famous corollary: The entropy of the universe tends toward a maximum. From this theorem many a philosopher maintained the conclusion of the impossibility of a world in which physical and chemical changes would go on being produced forever; it pleased them to think that these changes had had a beginning and would have an end; creation in time, if not of matter, at least of its aptitude for change, and the establishment in a more or less remote future of a state of absolute rest and universal death were for these thinkers inevitable consequences of the principles of thermodynamics.

The deduction here in wishing to pass from the premises to these conclusions is marred in more than one place by fallacies. First of all, it implicitly assumes the assimilation of the universe to a finite collection of bodies isolated in a space absolutely void of matter; and this assimilation exposes one to many doubts. Once this assimilation is admitted, it is true that the entropy of the universe has to increase endlessly, but it does not impose any lower or upper limit on this entropy; nothing then would stop this magnitude from varying from $-\infty$ to $+\infty$ while the time itself varied from $-\infty$ to $+\infty$; then the allegedly demonstrated impossibilities regarding an eternal life for the universe would vanish. But let us confess these criticisms wrong; they prove that the demonstration taken as an example is not conclusive, but do not prove the radical impossibility of constructing a conclusive example which would tend toward an analogous end. The objection we shall make against it is quite different in nature and import: basing our argument on the very essence of physical theory, we shall show that it is absurd to question this theory for information concerning events which might have happened in an extremely remote past, and absurd to demand of it predictions of events a very long way off.

What is a physical theory? A group of mathematical propositions whose consequences are to represent the data of experiment; the validity of a theory is measured by the number of experimental laws it represents and by the degree of precision with which it represents them; if two different theories represent the same facts with the same degree of approximation, physical method considers them as having absolutely the same validity; it does not have the right to dictate our choice between these two equivalent theories and is bound to leave us free. No doubt the physicist will choose between these logically equivalent theories, but the motives which will dictate his choice will be considerations of elegance, simplicity, and convenience, and grounds of suitability which are essentially subjective, contingent, and variable with time, with schools, and with persons; as serious as these motives may be in certain cases, they will never be of a nature that necessitates adhering to one of the two theories and rejecting the other, for only the discovery of a fact that would be represented by one of the theories, and not by the other, would result in a forced option.

Thus the law of attraction in the inverse ratio of the square of the distance, proposed by Newton, represents with admirable precision all the heavenly motion we can observe. However, for the inverse square of the distance we could substitute some other function of the distance in an infinity of ways so that some new celestial mechanics represented all our astronomical observations with the same precision as the old one. The principles of experimental method would compel us to attribute exactly the same logical validity to both these different celestial mechanics. This does not mean that astronomers would not keep the Newtonian law of attraction in preference to the new law, but they would keep it on account of the exceptional mathematical properties offered by the inverse square of the distance in favor of the simplicity and elegance that these properties introduced into their calculations. Of course, these motives would be good to follow; yet they would constitute nothing decisive or definitive, and would be of no weight the day when a phenomenon would be discovered which the Newtonian law of attraction would be inept to represent and of which another celestial mechanics would give a satisfactory representation; on that day astronomers would be bound to prefer the new theory to the old one.

That being understood, let us suppose we have two systems of celestial mechanics, different from the mathematical point of view, but representing with an equal degree of approximation all the astronomical observations made until now. Let us go further: let us use these two celestial mechanics to calculate the motions of heavenly bodies in the future; let us assume that the results of one of the calculations are so close to those of the other that the deviation between the two positions they assign to the same heavenly body is less than the experimental errors even at the end of a thousand or even ten thousand years. Then we have here two systems of celestial mechanics which we are bound to regard as logically equivalent; no reason exists compelling us to prefer one to the other, and what is more, at the end of a thousand or ten thousand years, men will still have to weigh them equally and hold their choice in suspense.

It is clear that the predictions from both these theories will merit equal degrees of confidence; it is clear that logic does not give us any right to assert that the predictions of the first theory, but not those of the second theory, will be in conformity with reality.

In truth these predictions agree perfectly for a lapse of a thousand or ten thousand years, but the mathematicians warn us that we should be rash to conclude from this that this agreement will last forever, and by concrete examples they show us to what errors this illegitimate extrapolation could lead us. The predictions of our two systems of celestial mechanics would be peculiarly discordant if we asked these two theories to describe for us the state of the heavens at the end of ten million years; one of them might tell us that the planets at that time would still describe orbits scarcely different from those they describe at present; the other, however, might very well claim that all the bodies of the solar system will then be united into a single mass, or else that they will be dispersed in space at enormous distances from one another. Of these two forecasts, one proclaiming the stability of the solar system and the other its instability, which shall we believe? The one, no doubt, which will best fit our extra-scientific preoccupations and predilections; but certainly the logic of the physical sciences will not provide us with any fully convincing argument to defend our choice against an attacking party and impose it on him.

So it goes with any long-term prediction. We possess a thermodynamics which represents very well a multitude of experimental laws, and it tells us that the entropy of an isolated system increases eternally. We could without difficulty construct a new thermodynamics which would represent as well as the old thermodynamics the experimental laws known until now, and whose predictions would go along in agreement with those of the old thermodynamics for ten thousand years; and yet, this new thermodynamics might tell us that the entropy of the universe after increasing for a period of 100 million years will decrease over a new period of 100 million years in order to increase again in an eternal cycle.

By its very essence experimental science is incapable of predicting the end of the world as well as of asserting its perpetual activity. Only a gross misconception of its scope could have claimed for it the proof of a dogma affirmed by our faith.

. . . .

9. ON THE ANALOGY BETWEEN PHYSICAL THEORY AND ARISTOTELIAN COSMOLOGY

Before proceeding further, let us summarize what we have gained above:

Between the ideal forms toward which physical theory and cosmology slowly travel, there ought to be an analogy. This assertion is by no means a consequence of positive method; although it is imposed on the physicist, it is essentially an assertion of metaphysics.

The intellectual procedure through which we judge the more or less broad analogy existing between a physical theory and a cosmological doctrine is quite distinct from the method through which convincing demonstrations are developed; they do not impose themselves.

This analogy should connect natural philosophy not to the present state of physical theory but to the ideal form toward which it tends. Now, this ideal state is not given in a plain and indisputable manner; it is hinted to us by an infinitely delicate and volatile intuition, whereas the analogy is guided by a profound knowledge of theory and its history.

The sorts of information which the philosopher can obtain from physical theory, either in favor of or against a cosmological doctrine, are therefore scarcely outlined indications; he would be very foolish who would take them as certain scientific demonstrations and be astonished to see them discussed and disputed!

After having thus definitely affirmed how much any comparison between a physical theory and a cosmological demonstration differs from a demonstration proper, after having indicated that it leaves plenty of room for hesitation and doubt, we shall be permitted to indicate the present form of physical theory which appears to us to tend toward the ideal form, and the cosmological doctrine which seems to us to have the strongest analogy with this theory. We do not maintain that this indication is to be given in the name of the positive method belonging to the physical sciences; after what we have said, it is obviously clear that it goes beyond the scope of this method, and that this method can neither confirm nor contradict it. In so doing, in penetrating thereby the domain belonging to metaphysics, we know that we have left the domain of physics behind us; we know that the physicist, after having gone along with us through the latter domain, may very well refuse to follow us into the terrain of metaphysics without violating logically imposed rules.

Which among the various ways, unequally favored by men of science, of dealing with physical theory at present is the one carrying the germs of the ideal theory? Which one already offers us through the order in which it arranges experimental laws something like a sketch of a natural classification? This theory, we have very often said, is in our opinion the one called general thermodynamics.

This judgment is dictated to us by the contemplation of the present state of physics and by the harmonious whole formed by general thermodynamics out of the laws discovered and made precise by experimenters; it is dictated to us, above all, by the history of the evolution which has led physical theory to its present state.

The movement through which physics has evolved may actually be decomposed into two other movements which are constantly superimposed on one another. One of the movements is a series of perpetual alternations in which one theory arises, dominates science for a moment, then collapses to be replaced by another theory. The other movement is a continual progress through which we see created across the ages a constantly more ample and more precise mathematical representation of the inanimate world disclosed to us by experiment.

Now, these ephemeral triumphs followed by sudden collapses making up the first of these two movements are the successes and reverses which have been experienced by the various mechanistic physical systems in successive roles, including the Newtonian physics as well as the Cartesian and atomistic physics. On the other hand, the continual progress constituting the second movement has resulted in general thermodynamics; in it all the legitimate and fruitful tendencies of previous theories have come to converge. Clearly, this is the starting point, at the time we live in, for the forward march which will lead theory toward its ideal goal.

Is there a cosmology which may be analogous to this ideal we glimpse at the end of the road where general thermodynamics engages physical theory? Surely it is not the ancient cosmology of the atomists any more than it is the natural philosophy created by Descartes, or the doctrine of Boscovich inspired by the ideas of Newton. On the contrary, it is a cosmology to which general thermodynamics is unmistakably analogous. This cosmology is the Aristotelian physics; and this analogy is all the more striking for being less anticipated and for the fact that all the creators of thermodynamics were strangers to Aristotle's philosophy.

The analogy between general thermodynamics and the physics of the Aristotelian school is marked by many a characteristic whose prominence attracts one's attention from the start.

Among the attributes of substance, equal importance is conferred by Aristotelian physics to the categories of quantity and quality; now, through its numerical symbols, general thermodynamics represents the various magnitudes of quantities and the various intensities of qualities as well.

Local motion was for Aristotle only one of the forms of general motion, whereas the Cartesian, atomistic, and Newtonian cosmologies agree in that the only motion possible is change of place in space. And notice that general thermodynamics deals in its formulas with a host of modifications such as variations in temperature or changes in electrical or magnetic state without in the least seeking to reduce these variations to local motion.

Aristotelian physics is acquainted with transformations still deeper than those for which it reserves the name of motions. Motion reaches only attributes; those transformations, viz., generation and corruption, penetrate to substance itself, creating a new substance at the same time that they annihilate a preexistent substance. Likewise, in the mechanics of chemistry, one of the most important chapters of general thermodynamics, we represent different bodies by masses which a chemical reaction may create or annihilate; within the mass of a compound body the masses of the components subsist only potentially.

These features, and many others, that it would take too long to enumerate, strongly connect general thermodynamics with the essential doctrines of Aristotelian physics.

We say "with the essential doctrines of Aristotelian physics," and we must now emphasize this point.

Experimental science was in its infancy at the time when Aristotle built the impressive monument whose plan has been conserved for us in his *Physics, On Generation and Corruption, On the Heavens,* and *Meteors*; and at the same time when his commentators, like Alexander of Aphrodisias, Themistius, Simplicius, Averroes, and innumerable scholastics, strove to chisel down and polish even the slightest portion of this enormous structure. The instruments which so greatly increase the extent, certainty, and precision of our means of knowing were not available to grasp material reality; man had only his naked senses; observable data came to him just as they appear first of all to our perception; no analysis had yet recognized and disentangled a frightful complication: facts, which a more advanced science was to consider as the results of a multitude of simultaneous, interlocked phenomena, were naïvely and hastily taken as the simple and elementary data of natural philosophy. The mark of everything which was incomplete, premature, and childish in this experimental science is necessarily in the cosmology which issues from it. One who hastily runs through the works of the Aristotelians and barely touches the surface of the doctrines expounded in these works notices everywhere strange observations, unimportant explanations, idle and fastidious discussions, in a word, an antique, worn out, deteriorated system in striking contrast with physics at present, so that it is only very remotely possible to recognize in them the slightest analogy with our modern theories.

Quite another impression is experienced by one who digs further. Under this superficial crust in which are conserved the dead and fossilized doctrines of former ages, he discovers the profound thoughts which are at the very heart of the Aristotelian cosmology. Rid of the covering bark which concealed them and at the same time held them in, those thoughts take on new life and movement; as they gradually become animated we see the mask of deterioration which disguised them disappear; soon their rejuvenated look and our general thermodynamics take on a striking resemblance.

He, then, who wishes to recognize the analogy of Aristotelian cosmology with theoretical physics today must not stop at the superficial form of this cosmology, but must penetrate to its deeper meaning.

An illustration may be brought in to clarify our thought and make it precise.

We shall borrow this illustration from one of the essential theories of Aristotle's cosmology, from the theory of the "natural place of the elements"; and we shall consider this theory on the surface, first of all, and, so to speak, from the outside.

In all bodies we always meet, although in various degrees, four qualities: the hot and the cold, the dry and the wet. Each of these qualities characterizes essentially one element: fire is eminently the hot element; air, the cold element; earth, the dry element; and water, the wet one. All the bodies surrounding us are mixtures; to the extent to which each of the four elements, fire, air, water, and earth, enter into the composition of a mixture, it is hot or cold, dry or wet. Beyond these four elements, capable of being transformed into one another by corruption and generation, there exists a fifth essence, incorruptible and nongenerative; this essence forms the celestial orbs and the stars which are condensed portions of these orbs.

Each of the elements has a "natural place"; it remains at rest when it is in this place, but when it is removed from it by "violence," it returns to it by a "natural motion."

Fire is essentially light; its natural place is the concavity of the moon's orb; by natural motion then it rises until it is stopped by this solid vault. Earth is the distinctively heavy element; its natural motion carries it to the center of the world which is its natural place. Air and water are heavy, but less heavy than earth; now, by natural motion the heavier trends to be placed below the lighter; the various elements will therefore be in their natural places when three spherical surfaces concentric with the universe separate water from earth, air from water, and fire from air. What maintains each element in its natural place when it is placed there? What carries it toward this place when it is removed from it? Its substantial form. Why? Because every being tends toward its perfection and in this natural place its substantial form attains its perfection; there it best resists anything which might corrupt it; there it experiences in the most favorable manner the influence of the celestial motions and astral light, the sources of all generation and of all corruption within sublunary bodies.

How childish all this theory of the heavy and the light seems to us! How plainly we recognize the first babblings of human reason trying to give an explanation of falling bodies! How dare we establish the slightest connection between these babblings of an infant cosmology and the admirable development of a science come to full vigor in the celestial mechanics of minds like those of Copernicus, Kepler, Newton, and Laplace?

Of course, no analogy appears between physics today and the first theory of natural place, if we take this theory as it appears at first sight with all the details making up its external form. But let us now remove these details and break this mold of outworn science into which the Aristotelian cosmology had to be poured; let us go to the bottom of this doctrine in order to grasp the metaphysical ideas which are its soul. What do we find truly essential in the theory of the natural place of the elements?

We find there the affirmation that a state can be conceived in which the order of the universe would be perfect, that this state would be a state of equilibrium for the world, and what is more, a state of stable equilibrium; removed from this state, the world would tend to return to it, and all natural motions, all those produced among bodies without any intervention of an animated mover, would be produced by the following cause: they would all aim at leading the universe to this ideal state of equilibrium so that this final cause would be at the same time their efficient cause.

Now, opposite this metaphysics, physical theory stands, and here is what it teaches us:

If we conceive a set of inanimate bodies which we suppose removed from the influence of any external body, each state of this set corresponds to a certain value of its entropy; in a certain state, this entropy of the set would have a value greater than in any other state; this state of maximum entropy would be in a state of equilibrium; all motions and all phenomena produced within this isolated system make its entropy increase; they therefore all tend to lead this system to its state of equilibrium.

And now, how can we not recognize a striking analogy between Aristotle's cosmology reduced to its essential affirmations and the teachings of thermodynamics?

We might multiply comparisons of this kind, and they would authorize, we believe, the following conclusion: If we rid the physics of Aristotle and of Scholasticism of the outworn and demoded scientific clothing covering it, and if we bring out in its vigorous and harmonious nakedness the living flesh of this cosmology, we would be struck by its resemblance to our modern physical theory; we recognize in these two doctrines two pictures of the same ontological order, distinct because they are each taken from a different point of view, but in no way discordant.

It will be said that a physics whose analogy with the cosmolgy of Aristotle and Scholasticism is so clearly indicated, is the physics of a believer. Why? Is there anything in the cosmology of Aristotle and in that of Scholasticism which implies a necessary adherence to Catholic dogma? May not a nonbeliever as well as a believer adopt this doctrine? And, in fact, was it not taught by pagans, by

Moslems, by Jews, and by heretics as well as by the faithful children of the Church? Where then is there that essentially Catholic character with which it is said to be stamped? Is it in the fact that a great number of Catholic doctors, some of the most eminent ones, have worked for its progress? In the fact that a Pope not long ago proclaimed the services that the philosophy of Saint Thomas Aquinas formerly rendered science as well as those that it may render it in the future? Does it follow from these facts that the nonbeliever cannot, without subscribing to a faith not his own, recognize the agreement of Scholastic cosmology with modern physics? Certainly not. The only conclusion that these facts impose is that the Catholic Church has on many occasions helped powerfully and that it still helps energetically to maintain human reason on the right road, even when this reason strives for the discovery of truths of a natural order. Now, what impartial and enlightened mind would dare to testify falsely against this affirmation?

10

The Slow Evolution of Science

Preface to Volume I of *Les Origines de la statique* (Paris: Hermann, 1905)
1905–11. Quoted, with permission, from the English translation, *The Origins of Statics*, by G. F. Leneaux, V. N. Valiente, and G. H. Wagener (Dordrecht: Kluwer Academic Publishers, 1991), vol. 1, pp. 7–9.

In this work the reader will not find the order which he undoubtedly would have preferred to find and which we assuredly would have preferred to follow. Therefore, we owe an explanation at the outset to the perplexed reader concerning our singular way of proceeding, in which our arguments might seem at times repetitive.

Before we embarked on the study of the origins of statics, we read the works — few in number — which deal with the history of this science. It was easy for us to see that most of these works were rather superficial and contained few details, but we had no reason to believe that this information was in any basic way incorrect. In resuming the study of the original texts mentioned in these works, we foresaw that we would have to add or modify quite a few details, but little did we know that our research would lead to a complete rethinking of the entire history of statics.

At the very outset, this research led us to make some unforeseen observations. It proved to us that the works of Leonardo da Vinci, so rich in new ideas on mechanics, had in no way remained unknown to the mechanicians of the Renaissance, as was commonly assumed. It further proved that his works were used by many scientists of the 16th century, in particular, by Cardan and Benedetti, and that they furnished Cardan with his profound insights on the operation of machines and on the impossibility of perpetual motion. However, after Leonardo and Cardan and up to Descartes and Torricelli, it seemed to us that statics developed along a path essentially in accordance with the commonly held views.

We had already commenced retracing this development in the *Revue des questions scientifiques*, when we chanced upon a text by Tartaglia, nowhere mentioned in any history of statics, which proved to us that what we had done so far had to be rethought on an entirely different level.

Indeed, it was Tartaglia who, long before Stevin and Galileo, had determined the apparent weight of a body on an inclined plane. He had very correctly deduced this law from a principle which Descartes was later to affirm in its complete generality. But this magnificent discovery, which no historian of mechanics mentions, did not come from Tartaglia. It was nothing but an impudent act of plagiarism on his part, and Ferrari bitterly reproached him for it and gave credit for this discovery to a 13th-century mechanician, Jordanus Nemorarius.

Two treatises had been published in the 16th century expounding the statics of Jordanus. However, these treatises were so dissimilar and sometimes contradicted each other so explicitly that they could not possibly be the work of the same author. In order to determine precisely what mechanics owed to Jordanus and his students, we had to go back to the contemporary sources, to the manuscripts.

Thus we were forced to go through all of the manuscripts dealing with statics which we were able to find at the Bibliothèque Mazarine. This laborious work, in which we were very competently assisted by M. E. Bouvy, librarian at the university of Bordeaux, led to totally unforeseen conclusions.

Not only did the Occidental Middle Ages directly or, indirectly through Arab intermediaries, inherit the tradition of certain Hellenic theories concerning the lever and the Roman balance, but through its own intellectual activity gave birth to a statics autonomous from and unknown in Antiquity. As early as the beginnings of the 13th century, and perhaps even earlier, Jordanus de Nemore had demonstrated the law of the lever by proceeding from the following postulate: the same work is needed to lift different weights when the weights are in inverse proportion to the heights which they travel through.

This idea, which can be found in germinal form in the treatise of Jordanus, was progressively developed in the works of his followers up to Leonardo da Vinci, Cardan, Roberval, Descartes, and Wallis, and reached its final formulation in the letter which John Bernoulli sent to Varignon, as well as in the *Mécanique analytique* of Lagrange and in the works of Willard Gibbs. Thus, the science which we are legitimately so proud of today grew out of a science born around the year 1300 through an evolution whose successive phases it is our task to describe.

However, it is not only through the doctrines of the School of Jordanus that the mechanics of the Middle Ages contributed to the foundation of modern mechanics. In the middle of the 14th century, Albert of Saxony, one of the doctors who brought great honor to the brilliant Nominalist School at the Sorbonne, formulated a theory on the center of gravity, which was to gain great recognition and have lasting influence. Shamelessly plagiarized during the 15th and 16th centuries by a great number of mechanicians and physicists who used his theory without naming its author, the theory continued to flourish during the 17th century. Indeed, to anyone ignorant of this fact, many a scientific controversy so hotly debated at that time would remain incomprehensible. Through an uninterrupted filiation, the principle of statics proclaimed by Torricelli developed from this theory of Albert of Saxony.

Thus the study of the origins of statics led us to a conclusion which became overwhelmingly evident when more varied avenues opened as we looked back in time. Therefore, only now do we dare formulate this conclusion in its full generality: the mechanical and physical science of which the present day is so proud comes to us through an uninterrupted sequence of almost imperceptible refinements from the doctrines professed within the Schools of the

Middle Ages. The so-called intellectual revolutions consisted, in most cases, of nothing but an evolution developing over long periods of time. The so-called Renaissances were frequently nothing but unjust and sterile reactions. Finally, respect for tradition is an essential condition for all scientific progress.

Bordeaux, March 21, 1905

11

Science and Providence

Conclusion of Volume II of *Les Origines de la statique* (Paris: Hermann, 1906) **1906-3**. Quoted, with permission, from the English translation, *The Origins of Statics*, by G. F. Leneaux, V. N. Valiente, and G. H. Wagener (Dordrecht: Kluwer Academic Publishers, 1991), vol. 2, pp. 438–448.

After the traveller has crossed the arid limestone plateau of Larzac with its grayish rounded hillocks and its maze of rocks resembling the ruins of a deserted city, he approaches the plain washed by the Mediterranean. The path which he must now follow is formed by steep ravines which are the traces of ancient streams or dried-out riverbeds and which, with the passing of time, cut deeper and deeper into the limestone plateau. Soon these ravines join in a single defile. Sheer walls, topped with ominous parapets of crumbling stones, now border the bed where once a beautiful river flowed deep and wild. Today, this riverbed is nothing but a chaos of worn and broken blocks. No spring flows from its rocky walls and no pool of water wets its gravel. Among the mass of stones, no plant can grow. The Vissec is the name given to the parched river of death by those who live here in the Cévennes.

The traveller, who can only advance with extreme exertion through the mass of fallen stones, occasionally hears a faraway rumble, like the roll of distant thunder. As he presses on, this rumble grows louder, before it bursts forth with a loud crashing sound. This is the great voice of the Foux.

In the limestone wall, a dark gaping cavern opens up like a giant maw. From this maw, white torrents of water gush forth and fall back thunderously in a mixture of crystal-clear droplets and bubbling white foam. The fissures in the distant limestone plateau have gathered this water in an underground lake.

Suddenly a river is born and henceforth, the clear and cold waters of the Vis flow between the white shores and the silvery oyster-beds. Its cheerful murmuring arouses in response the clicking of the mills and the deep ringing laughter of the Cévennes villages, while a glorious sunbeam glides over the notched edge of the plateau and slips down towards the bottom of the gorge, painting the poplar branches with a gold lining.

When traditional history, falsified by prejudice and distorted by deliberate simplifications, attempts to recount the development of the exact sciences, the image which it calls to mind resembles that of the course of the Vis River.

In this traditional view, Greek science flooded vast areas with its abundant and fertile waters. At that time the world witnessed the great discoveries of the likes of Aristotle and Archimedes germinate and grow and they will be admired forever. Then the source of Greek thought dried up and the river to which it had given birth ceased to deliver its life-giving force to the Middle Ages. The barbaric science of those days was nothing but chaos in which the unrecognizeable remnants of Hellenic wisdom were piled up helter-skelter, arid and sterile scraps to which clung, like parasitical, gnawing lichen, the puerile and conceited glosses of the commentators. All of a sudden, a great clamour shook this scholastic desert. Powerful minds cut through the rocks which had hidden for centuries the pure water from the ancient sources. Set free by these efforts, the waters gushed forth, happily and abundantly. Wherever they flowed, they brought about the rebirth of the sciences, of literature and the arts. The human mind simultaneously recovered both its vitality and its freedom. Soon thereafter, the great doctrines were born which, as the centuries passed, grew ever deepening roots and spread ever more impressively their branches and foliage.

But this traditional view is sheer nonsense. During its evolution, human science has had very few instances of sudden births or rebirths, just as the Foux is an exception among the sources of rivers.

A river cannot suddenly fill a large river bed. Before a river flows majestically, it is at first a mere stream, and a thousand other similar streams had to become its tributaries. At times those tributaries become numerous and abundant and the river rises rapidly. But when the tributaries are but few, mere trickles of water, the river hardly rises. Sometimes even, a cleft in the porous soil engulfs part of its waters and diminishes its flow. But through it all, its flow varies only gradually and does not disappear completely or spring out of nowhere.

Nor does science, in its progressive advance, have any sudden changes. It grows, but by increments. It advances, but by steps. No individual human intelligence, whatever its force or originality, will ever be able to produce a completely new doctrine at one stroke. The historian enamoured of simple and superficial views, delights in the brilliant discoveries which illumined the dark night of ignorance and error with the bright daylight of truth. Yet, anyone who is willing to analyse deeply and carefully what first appears to be a unique and unexpected discovery must soon conclude that it was the result of a great many imperceptible efforts and a conjunction of an infinite number of hidden tendencies. Each phase of the evolution which moves science slowly toward its goal has two characteristics: continuity and complexity. These characteristics are very clearly evident to the student of the origins of statics.

The historian enamoured of over-simplification will mention only one work produced on statics by the Ancients: the work of Archimedes. Such a historian will present the work as an isolated colossus, towering above the ignorance surrounding it. But to appreciate the greatness of this work, it is not necessary to distort it by presenting it in total isolation. The statics of the geometer of Syracuse is characterized by research of impeccable rigor in its deduction. His research is a subtle analysis applied to complicated problems and it presents wonderfully clever solutions to problems of which no one but a geometer understands the importance. All of those characteristics are the hallmark of a refined science and are far from the groping hesitations of a nascent doctrine.

It is obvious that Archimedes had precursors, who before him and by methods different from his own had understood the laws of the equilibrium of the lever which Archimedes was to develop so magnificently.

History has retained traces of some of these precursors. The *Mechanical Problems* may not be by Aristotle, as tradition would have it. In any case, the statics explained there is so closely connected with the dynamics set forth in the *Physics* and in the *On the Heavens* that we must attribute the *Mechanical Problems* to some close disciple of the Stagirite. The methods of a demonstration followed in this work might have been methods of invention. However, the same cannot be said about the deductions of Archimedes.

On the other hand, an ancient tradition stubbornly maintains that the writings about the lever are to be attributed to Euclid. These writings might not be those which we possess today under the name of the great geometer. Yet, merely denying their existence hardly proves the contrary.

If there were predecessors to Archimedes in Antiquity, he also most certainly had successors. The science of Byzantium and Alexandria followed the different paths outlined by Archimedes. The art of engineering brought to a high degree of perfection by the great Syracusan, inspired the works of Ctesibius, Philo of Byzantium and Hero of Alexandria.

Pappus, on the other hand, turned his attention to the study of centers of gravity, hoping in this fashion to become the equal of the Geometer. The enigmatic Chariston, finally, carried the principles of statics beyond Aristotle and Archimedes with his arguments on the Roman balance.

The Arabs transmitted only a tiny fraction of Hellenic statics to the Western medieval world, but they are far from being the servile and unimaginative commentators they are usually depicted. Their minds are quite receptive to the remnants of Greek thought which reached them via Byzantine and Islamic science. These remnants are enough to arouse their attention and to stimulate their minds. From the XIIIth century on, and perhaps even earlier, the School of Jordanus opens up avenues for the mechanicians which Antiquity had not known.

The intuitions of Jordanus de Nemore are initially rather vague and uncertain. Serious errors are mixed with great truths. But, little by little, the disciples of the great mathematician clarify the ideas of their master. Errors disappear and truths become more precise and established. Several of the most important laws of statics are finally confirmed with full certainty.

To the School of Jordanus we owe, in particular, a principle which will continue to grow in importance with the further development of statics. This principle bears no analogy to the special postulates on the lever upon which Archimedes had based his deductions, and it has only a slight affinity with the general axiom of Peripatetic dynamics. It asserts that the same motor force can lift different weights to different heights, provided that the heights are inversely proportional to the weights. Thanks to this principle, which Jordanus had applied exclusively to the straight lever, the Precursor of Leonardo da Vinci is able to grasp the law of equilibrium of the bent lever, the notion of moment, as well as the apparent weight of an object place on an inclined plane.

During the XIVth and XVth centuries, the statics which had originated in the School of Jordanus quietly pursues its own course and no other important stream of ideas adds to its flow. Not until the beginning of the XVIth century does it become a raging torrent, when Leonardo da Vinci brings to it his genius.

Leonardo da Vinci is far from being a seer who all at once discovers previously unsuspected truths. He is in possession of a prodigiously active mind, which is simultaneously daring and cautious. He returns to the laws of mechanics established by his predecessors, discusses them, and analyzes their every aspect. His relentless reflections enable him to refine several ideas already known to the disciples of Jordanus, and to demonstrate their richness and fecundity. An example of this is the notion of the motor force or the notion of moment. By means of a marvelous demonstration he is able to extract from this latter notion the law of the composition of concurrent forces. Yet, his mind, given to hesitation, alteration and reversal of opinion, does not always hold firmly to the truths which it has once grasped. Leonardo is unable to come to a definitive conclusion on the problem of the inclined plane, which had been quite satisfactorily resolved since the XIIIth century.

The indecision which always plagued Leonardo's soul and which so rarely allowed him to complete a work, kept him from bringing to completion the *Treatise on Weights* which he wanted to write. The fruits of his intellectual labors, however, were not entirely lost to science. Through the oral tradition which had originated during his lifetime, through the dispersion of his manuscripts after his death, his thoughts were scattered to the four winds and many fell on soil propitious to their growth.

Cardan, one of the most universal minds and also one of the most bizarre men produced by the XVIth century, and Tartaglia, a mathematical genius and a shameless plagiarist, both recovered for the statics of the Renaissance several discoveries made by the School of Jordanus. What they recovered was often in the richer and more fertile form that Leonardo da Vinci had given those discoveries.

Through the works of Tartaglia and Cardan a current of medieval mechanics starts to spread during the XVIth century. At the same time, a counter-current develops and gains strength in the treatises of Guido Ubaldo del Monte and Giovanbattista Benedetti. The works of Pappus and Archimedes had just been unearthed and were being studied and discussed with great passion and skill. They restored in mechanicians the desire for impeccable rigor, the hallmark of geometers since Euclid. This enthusiastic but narrow-minded admiration for the monuments of Hellenic science had only contempt for the profound but still flawed discoveries which the XIIIth century Schools had produced. The most profound intuitions of Jordanus and his disciples are misunderstood by the new school, which impoverishes and weakens statics while pretending to purify it. In the same spirit, the exclusive admiration for the works imbued with Hellenic grace scornfully rejects as "gothic" the most marvelous artistic creations of the Middle Ages.

Thus at the end of the XVIth century, almost nothing remained of what the specifically Western mind had produced in statics. Everything had to be done again. One had to redo the demonstrations of the truths which the medieval scholars had grasped and give them the clearness, precision and rigor of the theories bequeathed to us by the Greeks. This work of restoration, which will last to the middle of the XVIIth century, will be the task of the most brilliant geometers of Flanders, Italy and France. Yet despite the extraordinary talents of those men, much groping and many false starts were involved before the work came to completion!

A rigorous deduction must assume certain axioms. Where are the postulates to be found to form a solid foundation for statics? Those formulated by Archimedes are extremely restricted. They barely suffice to deal with the equilibrium of the straight lever. It becomes absolutely necessary to resort to new hypotheses. The mechanicians who are going to formulate them will present them as original principles and previously unknown truths. However, once we remove the pretense of originality, in which the vanity of those who had proclaimed them as truth had wrapped them, we are faced in almost every case with very old propositions, kept alive and nurtured by a long tradition which demonstrated their richness. Where a short-sighted and overly rigid historical view thought it could see a Renaissance of the scientific method which had fallen into oblivion since the Greeks, we ourselves see nothing but the natural evolution of medieval mechanics.

Galileo, as legend would have it, the creator of modern dynamics, returns to the already tottering dynamics of Aristotle to find the foundation for his deductions. He postulates the proportionality between the force moving a body and the velocity of that body. He is influenced by the mechanicians of the XIIIth century when he attempts to deduce from this principle the apparent weight of an object placed on an inclined plane. However, he fails to conclude from the works of these same mechanicians that the cardinal notion of all of statics is the notion of motor power, the product of a weight and the distance of its fall. Galileo replaces this notion with that of *momento*, the product of the weight and the velocity of its fall, a notion directly related to the previously rejected dynamics of Aristotle.

When Stevin deals with apparent weight on an inclined plane, he posits the impossibility of perpetual motion. However, Leonardo da Vinci and Cardan had already formulated this principle with remarkable clarity by linking it with the notion of motor power which they, in turn, had taken from the School of Jordanus. However, this notion plays only a secondary role in the work of Stevin. The great geometer from Brugge failed to recognize its extreme importance.

The same notion appears more clearly in the beautiful demonstration which Roberval furnished of the law describing the composition of concurrent forces. This demonstration, which fills so well the major deficiency in the work of Stevin, does not come out of nowhere. In order to deal with the equilibrium of the bent lever, the disciple of Jordanus who was the Precursor of Leonardo da Vinci, had already sketched the model for it.

The wonderfully insightful and systematic genius of Descartes soon grasped unerringly the main idea which must serve as the basis for all of statics. This idea is the same which Jordanus had already called attention to in his theory of the straight lever and is also the same idea used by his disciple in dealing with the bent lever and the inclined plane. It is the notion of motor power. Descartes defines this notion with great precision and shows its superiority to Galileo's use of *momento*. While the concept of momento derives from a dynamics henceforth untenable, the notion of motor power allows one to formulate a very clear, very certain axiom upon which all of statics can rest. To be accepted, this autonomous principle does not have to wait until the new dynamics has been built upon the ruins of Peripatetic dynamics.

Unfortunately, the obsessive arrogance which so often dominates Descartes causes him to exaggerate the magnitude of the

service which he renders to statics, and even to exaggerate to the
point of distorting the facts. Descartes, who was even less disposed
than Stevin, Galileo or Roberval to do justice to his predecessors,
portrays himself as the creator of a doctrine when he is merely its
organizer. What we have said here about Cartesian statics could
probably be said of all Cartesianism. The haughtiness of its author
has triumphed, and its triumph is unequalled in the history of the
human mind. It has duped the entire world and has portrayed
Cartesianism as a strangely spontaneous and unexpected creation.
However, the Cartesian system is in most cases nothing else but
the clearly formulated conclusion of anonymous efforts pursued
throughout centuries. The gracious flight of the butterfly showing
off its irridescent wings causes us to forget the slow and laborious
creeping of the humble and unobtrusive caterpillar.

The few lines in which Jordanus demonstrated the law of the
straight lever contained the seed of an accurate and fertile idea.
This idea continued to develop from Jordanus to Descartes until
it encompassed all of statics. While the gradual evolution of this
truth occurs, science becomes the scene of another equally
interesting, but more bizarre phenomenon. An incorrect doctrine
is slowly transformed into a very profound and correct principle.
It seems as if a mysterious force is watching over the progress of
statics and is able to render beneficial both truth and error alike.

Archimedes had used the notion of the center of gravity
without defining it. Several other geometers had attempted to
make it more precise. However, Albert of Saxony and, after him,
the majority of the physicists of the School, took advantage of
mechanical imprecision on this point, and attributed to it characteristics which are completely different from those which we
ascribe to it today. They considered, for example, that weight was
concentrated in each part of a body. The weight of a body
appeared to them as the desire of the center of gravity of that
body to join the center of the Universe. The Copernican revolution which relocated the center of the Universe and even went so
far with Giordano Bruno as to deny the existence of that center,
hardly modified this theory of weight. It considered this quality to

be the tendency inherent in the center of gravity of each body to join its analog, the center of gravity of the earth.

One of Kepler's claims to glory is to have eloquently fought against this hypothesis of an attraction between geometrical points and to have asserted that the attraction of gravity was exerted between different parts of the earth, taken two by two. However, his less clairvoyant contemporaries did not share this opinion. Benedetti, Guido Ubaldo, and Galileo in particular, postulated a sympathy felt by the center of gravity of each body towards the common center of heavy bodies, while Bernadino Baldi and Villalpand plagiarized the accurate corollaries which Leonardo da Vinci had deduced from this erroneous doctrine.

When this tendency is as completely satisfied as the connections of a system of weight allow it to be; when, in other words, the center of gravity of a system is as close as possible to the center of the earth, nothing induces this system to move any more and it remains in equilibrium. This is the principle of statics as formulated by Cardan, Bernardino Baldi, Mersenne, and Galileo and which perhaps they all borrowed from Leonardo da Vinci.

However, this principle is erroneous. In order to render it accurate, one needs only to extend to infinity that center of the earth which Galileo never ceases to invoke in his reasonings, and to consider the verticals as being parallel to each other. This modification seems insignificant. However, it is in truth profound, because it transforms a false statement into an accurate and fertile axiom. It is also profound because it presupposes the final abandonment of a very ancient and very authoritative theory of weight.

The confusing and complicated debates provoked in Florence by the research of Beaugrand and Fermat on the variation of weight according to its elevation prepare this reform and Torricelli completes it by giving science a new postulate upon which statics could be founded.

After having traced the continuous and complex development of statics, when the historian looks back to contemplate the total view offered by this science, he can only be astounded when he

compares the breadth of the finished theory with the minuteness
of the seed which brought it forth. On the one hand, he will be
able to decipher several lines of an almost illegible XIIIth century
manuscript in Gothic handwriting. Those lines justify very
precisely the law of equilibrium of a straight lever. On the other
hand, he can leaf through endless treatises of the XIXth century
in which the Method of Virtual Displacements helps to formulate
the laws of equilibrium for purely mechanical systems, as well as
for those where physical changes, chemical reactions, electric or
magnetic phenomena can occur. What a world of difference
between the simple demonstration of Jordanus and the imposing
doctrines of Lagrange, Gibbs, and Helmholtz! And yet, these latter
doctrines were contained potentially in that demonstration. History
has allowed us to retrace, step by step, the efforts through which
they were developed from that tiny seed.

The contrast between the extremely small and simple seed
and the strikingly complex, finished theory is analogous to what
the naturalist sees when he looks at the development of a plant or
an animal of a higher order. However, this marked contrast will
not excite his admiration as much as another spectacle far more
worthy of his attention and reflection. The process which he is
analyzing is the result of an infinite number of different phenomena. A great many cell divisions, budding transformations, and
reabsorptions are needed to produce the end result. All these
phenomena, however numerous, varied and complex they may be,
are coordinated with perfect precision. All of them combine very
efficiently for the formation of a plant or of an adult animal. And
yet, the countless beings which act in these phenomena, the cells
which proliferate, the phagocytes which devour the tissues which
have become useless, most certainly are unaware of the final goal
they are working to reach. They are workers ignorant of the final
product, but who, nevertheless, methodically bring to finality that
product. Thus the naturalist cannot help but seek outside of and
above these individual efforts something which is hard to define,
but which already has the final product of the plant or the animal
in mind and which during the formation of the organism sees to
it that the multitude of unconscious efforts combine to obtain the

end product. Like Claude Bernard, the naturalist will accept a "guiding idea" as presiding over the development of every living being.

Anyone who studies the history of science is led to similar reflections. Each proposition in statics was slowly elaborated through a process of research, experimentation, hesitations, discussions, and contradictions. Among all these many efforts, not one was wasted. Each one contributed to the final result. Each one played a greater or lesser role in the formation of the final doctrine. Even error proved fertile. The erroneous and sometimes bizarre ideas of Beaugrand and Fermat forced geometers to sift through the theory on the center of gravity in order to separate the precious truths from the falsehoods which had been intermingled.

Yet, while all these efforts contributed to the advance of a science which we can admire today in its finished form, no single contributor to these efforts even suspected the final magnitude and shape of the edifice he was helping to construct. When Jordanus developed the law of equilibrium for a straight lever, he was certainly not aware that he was formulating a principle which could form the basis for all of statics. Neither Bernoulli nor Lagrange had any inkling that their Method of Virtual Displacements would one day be perfectly suited to deal with electric and chemical equilibrium. They could not anticipate Gibbs, even though they were his predecessors. Like skillful masons cutting and cementing stone, they worked on the completion of an edifice without ever having seen the overall design of the architect.

How could all these efforts combine with such precision and bring to completion a plan which was not known to the individual laborer, unless this plan existed previously in the mind of an architect, and if this architect did not have the power to direct and coordinate the labor of the masons? Even more than the growth of a living being, the evolution of statics is the manifestation of the influence of a guiding idea. Within the complex data of this evolution, we can see the continuous action of a divine wisdom which foresees the ideal form towards which science must tend and we can sense the presence of a Power which causes the efforts of all thinkers to converge towards this goal. In a word, we recognize here the work of Providence.

Bordeaux, October 26, 1905.

12

The Coherence of Scientific Evolution

Preface to Volume I of *Etudes sur Léonard de Vinci* (Paris: Hermann, 1906), pp. iii–vii **(1906–10)**.

When we contemplate a great discovery, we experience first of all an admiration mixed with fright. Our astonished look gauges the height to which the genius is raised. We feel the extent to which that height surpasses all those that our humble mind could reach and a sort of vertigo seizes us.

Then, in the measure in which our meditation makes us more familiar with the discovery which astonished us, our admiration changes nature. Not that it loses its intensity; but it divests itself step by step of all instinctive and unreflexive elements which our surprise may have mixed into it. Our admiration becomes more and more conscious and reasoned. Colossal as the genius may appear to us, we realize that it is of a nature no different from our modest intelligence, that it proceeds by the same routes as does our mind, though with an incomparable sureness and promptness. We clearly see that it did not rise by one bound to the heights we contemplate, that it arrived there by a long series of efforts entirely similar to those of which we are capable. Then the desire is born in us to know each of those efforts and the order in which they follow one another. We want the detailed account of the ascent which led the inventor to his discovery.

But how difficult it is to obtain this account in an exact and precise form!

Very often the one who has arrived at the summit where he discovers a vast truth has no other concern than to describe to others the view available to him. As to the pains which he took to reach the peak, where his view can extend far and wide, he has forgotten them and takes them to be miseries with no importance, unworthy of being recounted. He gives us his completed work, but throws into the fire his sketches.

Others tell us how they happened to become discoverers but it is not always wise to trust their confessions.

From the top one can see all the roads that can lead there, but one did not suspect them until one climbed the mountainside. Among the roads one sees, on occasion, one which is more simple and easy, along which one could have avoided the long detours and the wrong steps. It is this easy route which the discoverer describes to us and not the painful and dangerous path which he had really followed. "My discovery is made," so goes a saying attributed to Gauss, "it remains for me to find out how I should have made it."

There are discoverers who hide from us with a sort of bashfulness the long and painful procedures of their minds in quest of truth. They show us only the royal road along which it would have been easy to make the discovery which cost them so much effort. There are others who boast of their vigor and skill. They tell us, or would have us believe, that they had guessed by their own talents alone most hidden paths, interrupted by risky passages. They are careful not to name the guide whose experience prevented them from going astray, and whose helpful hand saved them from falling. They describe to us with complaisance the complicated turns of their deductions and the profundity of their reflections. They do not tell us the readings that guided those deductions and supported those reflections.

It is therefore rather difficult to follow the progress of an idea in the mind of a discoverer and to develop the series of forms by which it has attained its perfection.

For our curiosity to be fully satisfied, it would be necessary for the discoverer to have carefully marked his advance in the measure in which he had accomplished it; that he had marked, so to speak, the trace of each of his steps. We would wish that each of his thoughts had been fixed in writing at the very moment when it took shape in his mind. Such notes, once collected, would allow us to understand how the idea became clarified bit by bit, from the moment when the genius had suspected its vague silhouette across the mist of doubts to the very moment where he could contemplate it in full evidence, in the splendor of its being a truth.

Now, among those who have initiated the human mind to the perception of new truths, there is one who left for us this meticulous description of the movements of his thoughts, who put in writing, so to speak, the diary of the voyage of discoveries that was his life. In the measure in which a new proposition offered itself to his reflections, he jotted it down with complete sincerity, without dissimulating any of his hesitations, any of his regrets, because he wrote only for himself. As a result his precious sketches allow us to follow, from the first sketches to the final and detail design, the diverse forms which a discovery had taken in the ingenious mind of Leonardo da Vinci.

The manuscripts of Leonardo da Vinci are documents of inestimable value because they are unique in their class. Not one of those whose reflections have enriched science have given us, on the subject of the march followed by his thoughts, indications so numerous, so detailed, and so direct.

Not that these documents give us right away and without a lengthy effort information which they contain in abundance.

Those short notes written with the left hand, difficult to decipher and often obscure in their extreme conciseness, are rarely dated. The notebooks which contain them were filled as much in the order of pagination as in the contrary sense. Some of these notebooks seem to contain reflections born at different phases of the great painter's life. Worse, a great number of them have been lost.

The task is to recover, from the midst of this chaos, diverse fragments that deal with the same discovery and put them in the temporal sequence in which they were conceived so that they would mark the successive steps of an idea in progress. This task is often very problematic and the results are not always of absolute certainty.

Cumbersome as is this task, it is perhaps not the most exacting that is to be accomplished in order to retrace the history of a discovery of Leonardo.

When a new idea is born in the mind of Leonardo, it does not appear there of itself and without cause. It is produced there by some external circumstance, by the observation of a natural phenomenon, by the words of a man, and, still more often, by the reading of a book.

At any rate, the mind into which fell this seed of thought was not at all similar to a barren and naked field. It had already been occupied by other thoughts, vigorous and pressing. They were implanted there by lectures of masters whom Leonardo had heard and especially by the teaching of writings he had meditated upon. In order to sprout and grow, it was necessary that the new seed be served by that vegetation already developed there or that it struggle against it.

If therefore one wants to follow the evolution of an idea in the mind of Leonardo da Vinci, one has, at the very start, to answer this question: "What did he read?" The reply cannot be given without long and meticulous researches. For one, by making his hasty and concise notes, Leonardo very rarely named the author whose lecture or memory had suggested to him such and such a proposition. For another, by comparing his work to those of his predecessors, one quickly discovers that he had read much and that he had studied a good number of the scientific tracts of his time.

One of the objectives of these studies is to make known some of the sources which Leonardo utilized and to establish what each of them contributed to the stream of the great discoverer's thoughts.

But in order to appraise exactly the role which Leonardo played as an initiator it is not enough to determine and to study *those whom he has read*. It is also necessary to discover those *who have read him*.

The idea on whose progress the discoverer worked does not acquire its fulness and completion in virtue of its author. When he published it, it is still pregnant with new truths. It will produce these new truths through the work of those who espouse the discovery and try to develop it. It is just to praise the first initiator not only for what he had actually put in his invention but also for what he had left there in potency. The latter cannot be studied except by studying the works of his successors.

Leonardo had his successors. His manuscript notes did not remain intact in a sterile oblivion. Impudently pillaged and plagiarized, they threw to the four corners of the sky seeds of truth which they contained in abundance. These seeds brought their fruits in the science of the 16th century. To name some of those who knew those notes, to expose the plagiarism, to evaluate what they owed to Leonardo is the second objective of these studies.

Between *those whom he had read* and *those who read him* Leonardo da Vinci appears in his true place. An integral part of the past whose teaching he had read and reflected upon, he is also an integral part of the future as one whose thoughts have fertilized science.

Bordeaux, July 27, 1906.

13

Evolution of Science and Christian Gratitude

Preface to Volume III of *Etudes sur Léonard de Vinci* (Paris: Hermann, 1913), pp. v–xiv **(1913–4)**.

To the third series of our *Etudes sur Léonard de Vinci* we give the subtitle: *Les précurseurs parisiens de Galilée*. This subtitle announces the idea of which our preceding studies had already unveiled some aspects and are now put in full light by our new studies. The science of mechanics inaugurated by Galileo, by his admirers, by his disciples — such as Baliani, Torricelli, Descartes, Beeckman, and Gassendi — is not a sudden creation. The modern mind did not produce it as if by a first move and in all its parts once the reading of Archimedes' writings revealed the art of applying geometry to natural phenomena. Galileo and his contemporaries used the mathematical skill developed in the interaction of the geometers of Antiquity to make more precise and further develop a science of mechanics for which the Christian Middle Ages laid the principles and formulated its most essential propositions. The physicists who taught this mechanics at the University of Paris during the 14th century conceived it by taking observation for guide. They have substituted it for the dynamics of Aristotle, convinced as they were of the latter's inability "to save the phenomena." During the Renaissance, that doctrine of the "moderns" was rejected by the superstitious archaism that delighted the mentality of the Humanists as well as by the routine averroism of retrograde scholastics. The reaction was a powerful, retrograde scholastics. The reaction was powerful, especially in Italy, against the dynamics of the "Parisians," and in favor of the inadmissible dynamics of the Stagerite. But in spite of this hard-headed resistance, the Parisian tradition found, outside the Schools as well as inside the universities, teachers and savants to maintain and develop it. It is this Parisian tradition of which Galileo and his admirers were the heirs. When we see the science of Galileo triumph over the propped up peripatetism of a Cremonini, we, misinformed about the history of human thought, believe that we assist at the victory of a young modern science over a medieval philosophy frozen in its prejudices. In reality, we see the triumph, for a long time prepared, of a science born in Paris in the 14th century over the doctrines of Aristotle and Averroes, reinstated in honor by the Italian Renaissance.

No movement can last unless maintained by the continued action of the moving power, directly and immediately applied to the thing moved. Such is the axiom on which rests all the dynamics of Aristotle.

In conformity with this principle, the Stagerite wants to apply to the arrow, which continues to fly after it has left the arch, a moving force that keeps transporting it. He thinks to find that power in the air which has been given a push. It is this air, set in motion either by the hand or by the ballistic machine that supports and carries the projectile.

This hypothesis, which to us seems to push the improbabilty to the point of ridicule, seems to have been admitted almost unanimously by the physicists of Antiquity. Only one among them has gone clearly on record against it and he, whom chronology places in the last years of Greek philosophy, finds himself, on account of his Christian faith, almost completely separated from that philosophy. We have in mind John of Alexandria, also called Philoponus. After he had shown what is inadmissible in the peripatetic theory of motion, John Philoponus declares that the arrow continues moving without being in contact with any moving force, because the cord of the arch has generated in it an *energy* that plays the role of moving force.

Neither the last thinkers of Greece, nor the Muslim philosophers have as much as mentioned the teaching of this John, the Christian, for whom a Simplicius and an Averroes have but sarcastic contempt. The Christian Middle Ages, struck by a naive admiration for the newly discovered peripatetic science, first shared the disdain of the Greek and Muslim commentators for Philoponus' hypothesis. Saint Thomas Aquinas mentions that hypothesis only to caution against it those whom it might seduce.

But in the wake of the condemnations brought, in 1277, by the bishop of Paris, Etienne Tempier, against a large number of theses that supported Aristotle and his followers, there arises a grand movement which will liberate Christian thought from the yoke of peripatetism and neoplatonism and produce what the archaism of the Renaissance will call the science of the moderns.

William of Ockham attacks, with his customary vivacity, the theory of projectile motion proposed by Aristotle. At any rate, he is contented with destroying without constructing anything. But his criticisms restore to honor the doctrine of John Philoponus among some disciples of Duns Scotus. The energy, the moving force of which Philoponus had spoken, re-appears under the name of *impetus*. This hypothesis of *impetus*, imparted to the projectile by the hand or by the machine that launched it, is seized upon by a physicist of genius, Jean Buridan. He takes it, toward the middle of the 14th century, for the foundation of a dynamics with which "all the phenomena are in accord."

The role which the *impetus* plays in this dynamics of Buridan is exactly that which Galileo will attribute to the *impeto* or *momento*, Descartes to the *quantity of motion*, and, finally, Leibniz to the *live force*. So exact is that correspondence that in order to set forth in his *Academic Lessons* Galileo's dynamics, Torricelli often repeats the reasoning and almost the very words of Buridan.

This *impetus*, which would remain without change within the projectile if it were not constantly destroyed by the resistance of the medium and by the action of gravity contrary to the motion, this *impetus*, let us state it, Buridan takes, for equal velocities, for something proportional to the *quantity of prime matter* contained within the body. He conceives and describes this quantity in terms almost identical with those which Newton will use to define mass. With equal mass, the *impetus* is as much greater as is the velocity. Wisely, Buridan abstains from specifying further the relation between the magnitude of *impetus* and that of the velocity. More daring, Galileo and Descartes grant that this relation reduces itself to proportionality. They thus obtain for *impetus*, for the *quantity of motion*, an erroneous evaluation which Leibniz will correct.

As does the resistance of the medium, gravity too continually diminishes, and finally annihilates the *impetus* of the projectile launched upward because such a motion is contrary to the natural tendency of gravity. But in an object that falls, the motion conforms to the tendency of gravity. Therefore the *impetus* too continues increasing and the velocity, during the motion, must constantly increase. Such is, according to Buridan, the explication of acceleration observed in the fall of a body, an acceleration which Aristotle's science already knew but of which the Hellenistic, Arabic, and Christian commentators of the Stagerite gave unacceptable explanations.

This dynamics, as set forth by Buridan, presents in a purely qualitative, though always accurate, manner the truths which the notions of living force and of work allow us to formulate in quantitative language.

The philosopher from Bethune [John Buridan] is not the only one to profess this dynamics. His most brilliant disciples, Albert of Saxony and Nicole Oresme, make it known even to those who are not from the clergy.

When there is no resisting medium, when no natural tendency similar to gravity opposes the motion, the *impetus* keeps an invariable intensity. The projectile to which one imparted a translational or rotational motion continues moving indefinitely with an invariable velocity. It is in this form that the law of inertia arises in the mind of Buridan. It is in this form that it will be inherited by Galileo.

From this law of inertia Buridan draws a corollary whose novelty we must now admire.

If the celestial orbs move eternally with a constant velocity, they do so because, according to an axiom of Aristotle's dynamics, each of them is subject to an eternal mover of unchangeable power. The philosophy of the Stagerite requires that such a mover be an intellect separate from matter. The study of intellects moving the celestial orbs is not only the crowning of the peripatetic metaphysics; it is also the central doctrine around which turn all the neoplatonist metaphysics of the Hellenist and Arabic philosophers. The Scholastics of the 13th century do not hesitate to accept, in their Christian systems, this heritage of pagan theologies.

It is right here that Buridan has the audacity to write these lines:

"At the creation of the world, God moved the heavens with motions identical to the ones with which they actually move. Therefore He then imparted to them *impetuses* by which they continue to move uniformly. Not encountering any resistance contrary to them, these *impetuses* are never diminished nor destroyed. . . . Within this view it is not necessary to posit the existence of intellects that move the celestial bodies in an appropriate manner."

This idea is stated by Buridan in various contexts. Albert of Saxony lectures on it and Nicole Oresme, in formulating it, finds this analogy: "Apart from the effort, this is altogether similar to the procedure when a man constructs a clock and lets it move itself."

If one wants to separate by an exact line the domain of ancient science from that of modern science, it has to be drawn, we believe, at the moment when John Buridan conceived that theory, at the very moment when one ceased to see the stars as if moved by divine beings, at the moment when one admitted that the celestial motions and the sublunary motions rested on the same mechanics.

Here is a mechanics, celestial and terrestrial at the same time, to which Newton would one day give the form we admire today, as it tries to develop from the 14th century on. During that century, as evidenced by François de Meyronnes and Albert of Saxony, there were physicists who claimed that by assuming the earth to be in motion one could construct an astronomical theory more satisfactory than the one in which the earth is deprived of motion. Among these physicists Nicole Oresme develops the reasons of this new mechanics with a fulness, clarity, and precision which Copernicus is far from displaying. Oresme attributes to the earth an *impetus* similar to the one which Buridan attributes to the celestial orbs. To account for the vertical fall of gravitating bodies, Oresme admits that one has to compose that *impetus* by which the body goes around with the earth with the *impetus* generated by gravity. The principle which Oresme clearly formulates Copernicus merely indicates in an obscure manner and Giordano Bruno simply repeats. Galileo will rely on geometry to draw its consequences without, however, correcting the erroneous form of the law of inertia which he finds implied in it.

While the foundations of the dynamics are established, step by step the laws that govern the fall of bodies are also discovered.

In 1368 Albert of Saxony proposes two hypotheses: the velocity of the fall is proportional to the time elapsed since its start, and, the velocity of the fall is proportional to the space traversed. Between these two laws he makes no choice. The theologian, Petrus Tartaret, who teaches in Paris toward the end of the 15th century, repeats word for word what Albert of Saxony had said. As a sedulous reader of Albert of Saxony, Leonardo da Vinci opts, after having favored the second hypothesis, for the first. Still, he does succeed in discovering the law of spaces traversed by a body that freely falls. Through a reasoning, which is reported by Baliani, Leonardo concludes that the spaces traversed in equal and successive times are like the series of integers, whereas, in truth, they are like the series of odd numbers.

For a long time, however, had been known the rule that allows the evaluation of the space traversed, within a certain time, by a body moved by a motion uniformly varied. That this rule had been discovered in Paris in Buridan's time or in Oxford in Swineshead's time is clearly stated in Nicole Oresme's work where he lays down the essential principles of analytical geometry. Moreover, the demonstration that is used there to justify them is identical to the ones to be given by Galileo.

From the time of Nicole Oresme to that of Leonardo da Vinci that rule was in no way forgotten. Formulated for the most part in treatises produced in terms of the intricate dialectic cultivated in Oxford, the rule is discussed, during the 15th century in Italy, in numerous commentaries that had that rule for their subject, and subsequently in the various works on physics composed by the Parisian scholasticism in the beginning of the 16th century.

None of these treatises we have just been talking about contains the idea of applying that rule to the fall of bodies. We encounter this idea for the first time in the *Questions on the Physics of Aristotle* published in 1545 by Dominic Soto. Disciple of the Parisian scholastics, whose guest he was and whose physical theories he adopted for the most part, the Spanish Dominican admits that the fall of bodies is uniformly accelerated, that the vertical ascent of a projectile is uniformly retarded. In order to calculate the spaces traversed by each of these motions, he makes a correct use of the rule formulated by Oresme. This means that he knew the laws of the fall of bodies, laws taken for a discovery of Galileo.

At any rate, he does not claim to himself the discovery of these laws. Rather, he seems to give them as truths generally accepted. Without doubt they had been routinely accepted by the teachers whose lectures Soto attended in Paris. Thus, from William of Ockham to Dominic Soto we see the physicists of the Parisian school lay down all the foundations of the mechanics which will be developed by Galileo, his contemporaries, and his disciples.

Among those who have received, prior to Galileo, the tradition of the Parisian scholasticism, there is none who merits greater attention than Leonardo da Vinci. At the time when he lived, Italy posed a firm resistance to the penetration of the mechanics of the "moderns," and of the "juniors." In Italy even those among the university teachers who leaned towards the teaching of the Parisian terminalists, limited themselves to repeating, under a short and hesitant form, the essential statements of that mechanics. They fell far short of producing any of the fruits of which that mechanics was the flower.

Leonardo da Vinci was not, however, satisfied with admitting the general principles of the dynamics of *impetus*. He constantly meditated on those principles and, looking at them from all angles, he, so to speak, forced them to reveal the consequences they contained. The essential hypothesis of that dynamics was something like the first form of the law of living force. Leonardo perceived there the idea of the conservation of energy and in order to express that idea he finds terms of prophetic clarity. Between two laws of the fall of bodies, one exact, the other inadmissible, Albert of Saxony left his reader in suspense. After some hesitation which Galileo will be familiar with, Leonardo knows how to fix his choice on the right law. He felicitously extends it to the fall of bodies along an inclined plane. Through the study of a composite *impetus* he is the first to try the explanation of the curvilinear trajectory of projectiles, an explanation which receives its completion from Galileo and Torricelli. Leonardo perceives the correction to be given to the law of inertia stated by Buridan and prepares the work to be accomplished by Benedetti and Descartes.

Without doubt, Leonardo does not always recognize all the richness of the treasury accumulated by the Parisian scholasticism. He ignores some whose contribution might have given to his doctrine on mechanics a most felicitous complement. He misjudges the role which the *impetus* was to play in the explanation of the fall of bodies. He ignores the rule that allows one to calculate the spaces traversed by a body moved by a uniformly accelerated motion. It is true, nevertheless, that his physics as a whole puts Leonardo among those whom the Italians of his time call Parisians.

At any rate, this title is justly given to him. He draws the principles of his physics from his assiduous reading of Albert of Saxony, and probably also from his meditations on the writings of Nicolas of Cusa, who himself was a follower of the Parisian mechanicians. Leonardo has therefore his place among the Parisian forerunners of Galileo.

Until these very recent years, the science of the Middle Ages had been taken for non-existent. A philosopher who admirably knew the history of science in Antiquity and during modern times, recently wrote:

"Suppose that the art of printing had been invented two centuries earlier. It would have helped maintain orthodoxy and would have merely served to propagate, apart from the *Summa* of

Saint Thomas and some other works of that type, the bulls of excommunications and decrees of the Holy Office."★

Today, we believe, we are allowed to say:

"If the art of printing had been invented two centuries earlier, it would have served the publications, at the rate as they were written, of the works which, on the ruins of Aristotle's physics, had laid the foundations of a mechanics of which the moderns are justly proud."

This substitution of modern physics for Aristotle's physics was the result of long duration and extraordinary acumen.

That effort based itself in the most ancient and most celebrated of medieval universities, the University of Paris. How should a Parisian not be proud of this?

The most eminent promoters of that new physics were John Buridan from Picardy and Nicole Oresme from Normandy. How could a Frenchman not experience a legitimate pride on that account?

The new physics was the result of the resolute struggle which the University of Paris, true guardian at that time of Catholic orthodoxy, kept waging against peripatetic and neoplatonic paganism. How could a Christian not be grateful to God for this?

★G. Milhaud "Science grecque et Science moderne," in *Comptes rendus de l'Académie des Sciences morales et politiques*, 1904. — G. Milhaud, *Etudes sur la pensée scientifique chez les grecs et les modernes*, Paris, 1906, pp. 268-69.

14

Galileo's Debt to the Medievals

From the Conclusion of *Etudes sur Léonard de Vinci* (Paris: Hermann, 1913), pp. 582–583 **(1913–4)**.

Galileo knew the kinematics of the Oxford School and, in a most fortunate way, he was subject to its influence.

Did he also know the dynamics of Paris, that dynamics of John Buridan and Albert of Saxony with which his ideas often offer striking analogies?

In his youthful writings Galileo cites twice the Parisian doctors, *Doctores Parisienses*.

In his treatise, *De elementis* he tells us that "according to Aristotle whom the Parisian doctors followed," the volumes of elements follow a tenfold progression. This opinion is in fact set forth in detail and accepted by Themon, son of the Jew, in the sixth question of the first book of his *Meteora*.

The second citation is more precise. In his *De coelo* Galileo lists the authors according to whom the world could have existed since eternity. "This opinion," he says, "is of Saint Thomas . . . , of Scotus . . . , of Ockham . . . , and of the Parisian doctors in the first question of the eighth book of Physics" (*Doctorum Parisiensium 8 Phys. q. prima*).

Here we see that by that collective name, Parisian Doctors, Galileo does not designate, in a general and vague manner, a certain school, but, in a precise manner a certain well determined work.

TEXTS

Now we can recall that in the first question of the eighth book of his Physics, Albert of Saxony declares, in fact, that (apart from the teaching of faith) the world and its motion could have existed since eternity.

What is therefore that work, composed by the Parisian Doctors, where, à propos of a question relating to the *Meteora* one encounters the opinion that Themon had admitted it in his *Meteora*; who, in the first question of the eighth book of Physics, teaches exactly what Albert of Saxony had taught in the first question of the eighth book of his *Physica*? But this is what conspicuously is not open to any ambiguity. We know that work. It is a collection, published in Paris — and twice at that, in 1516 and 1518 — in which George Lockart assembled the *Physica,* the *De coelo*, the *De generatione et corruptione* of Albert of Saxony, the *Meteora* of Themon, and the *De anima* and *Parva naturalia* of John Buridan. It is this collection that Galileo had read at the time when he was writing his scholastic dissertations. It is through that collection that he became initiated in the dynamics of Paris.

Is it not therefore permissible now to invoke the testimony of the ingenious Pisan to salute these Parisian doctors as being the Forerunners of Galileo?

15

Two New Chairs for Catholic Universities

For the French original, see Hélène Pierre-Duhem, *Un savant français: Pierre Duhem* (Paris: Plon, 1936), pp. 158–169.

Bordeaux, May 21, 1991

Dear Father,

I have heard that the Institut Catholique de Paris is ready to put together a systematic set of courses for instructions in philosophical subjects. This news causes me great joy and will, I think, cause great joy to all clear-sighted Catholics. It is time that we oppose to the numerous and learned teachings in philosophy a school made up of chairs from which the traditional philosophy of Catholicism will be set forth in all its strength and development.

On the subject of the composition of the future Institut de Philosophie, some reflections have come to my mind and I beg your permission to share them with you. They are not counsels, which, were they come from me, would be impertinent; they are rather simple considerations. As one living among those who profess doctrines contrary to ours, I am well placed to know their plan of attack against us and to see where our defenses ought to be specially reinforced.

The field where the battle is already on, where, without doubt, it will become ever more violent, is the incompatibility of the scientific mind and the religious spirit.

I am not speaking of the incompatibility of a specific scientific discovery with specific religious doctrine. Much polemics was made about these particular antagonisms during the 19th century. Then it was fashionable, for instance, to oppose a theory in geology to a verse in the Bible. But these were isolated skirmishes that prepared the great engagement. This is of far greater amplitude and the target it aims at threatens to be much more radical. It amounts to denying, in the name of science as such, to all religion the very right to exist. They pretend to have established that no sensible man can admit, at the same time, the value of science and also believe in the dogmas of religion. And since the value of science asserts itself every day by thousands of marvelously useful inventions, a fact which only a blind mind can call into doubt, the religious faith is taken for a goner.

In order to establish that essential and absolute incompatibility between all science and all religion, an appeal is made to the logical analysis of the methods by which one and the other operate.

Science, they say, takes for its basis axioms that reason cannot deny or facts that have all the certainty of the testimony of the senses. All that rises on such foundations is constructed with the help of rigorous reasoning. Moreover, experiments with additional precautions control each of the conclusions to which science arrives. The entire edifice obtains thereby an unshakable solidity of prime foundations.

The religious dogmas, on the contrary, originate in vague and elusive intuitions and aspirations that are born of sentiments and not at all of reason. They are not submitted to any rule of logic and could not stand, even for a moment, an examination by a modestly rigorous criticism.

From that point on one either declares that the object of religious dogmas is absurd and void of sense, or one is contented with a narrow and rigid positivism, very akin to gross materialism which is its rigorous consequence. Or one considers religion as an object which escapes the demonstrations of science and is therefore incapable of being known with the slightest certitude. One will profess an agnosticism for which all religion is a dream, more or less poetical and comforting. But how will anyone who had experienced the firm realities of science let himself be lulled by such a dream?

This antagonism between the scientific spirit and the religious mind is being set forth not merely by the tools of logic. It is further desired that also the history of the development of human knowledge place in broad light this antagonism for the benefit of the less clearsighted. They show us how all the sciences are born of the fertile Greek philosophy whose most brilliant exponents left to the vulgar the ridiculous concern for believing in religious dogmas. They depict to us shockingly that night of the Middle Ages during which the Schools, subservient to the agencies of Christianity and exclusively concerned with theological discussions, did not know how to gather the smallest parcels of the scientific bequest of the Greeks. They make shine into our very eyes the glories of the Renaissance where minds, liberated at long last from the yoke of the Church, have found again the thread of scientific tradition at the same time as they found the secret of scientific and literary beauty. They delight in contrasting from the 16th century on the always ascending march of science with the ever deeper decadence of religion. They believe themselves to be authorized to prophesy the imminent demise of religion and at the same time the universal and unchallenged triumph of science.

This is what is being taught in a number of chairs, this is what is being written in a multitude of books.

In the face of that teaching it is time that Catholic teaching rise and hurl into the eyes of its adversary the word: Lie! Lie in the domain of logic, lie in the domain of history. A teaching which pretends to have established the irreducible antagonism between the scientific spirit and the spirit of Christianity is the most colossal lie and also the most audacious which has ever attempted to dupe the people.

In order to oppose the method that leads to scientific truths to the method that leads to religious dogmas, they describe falsely one and the other of these methods. They consider both in a superficial manner. They seize on some features revealed by their rapid investigation and make them even the essence of the procedure which they pretend to have analyzed.

How different those methods appear to the one who has penetrated them to the core and who has seized in each the principle of life. He can recognize at the same time what gives verity to these procedures and what unites them. Everywhere he sees the one and the same human reason make use of the same essential tools in order to arrive at the truth. But in each domain he sees that reason adapt the use it makes of these tools to the special object whose knowledge it wants to acquire. Thus, aided by common operations that constitute our intellect, he in turn sees it follow a method of mathematics, a method of physics, a method of chemistry, of biology, of sociology, of history; for mathematics, physics, chemistry, biology, sociology, history have different principles and different objects, and in order to attain those objects different starting points are necessary, and by the same token different routes are also to be followed. He then recognizes that in order to reach religious truths, the human reason uses no other tools than the ones it uses to attain other truths, but it uses them in a different manner because the principles from which it starts and the conclusions to which it tends are different. The antagonism which was decried to exist between scientific demonstration and religious intuition disappears from his eyes while he perceives the harmonious accord of multiple doctrines whereby our reason tries to express truths of different order.

What to say of the curious story with which they try to confirm what an insufficient logical analysis asserted with ease?

From its birth, Greek science is all impregnated with theology, but with a pagan theology. That theology teaches that the heavens and the stars are gods. It teaches that they cannot have other motion than circular and uniform motion which is the perfect motion. It curses the impiety that would dare to attribute a motion to the earth, sacred foyer of the divinity. If these theological doctrines have furnished some postulates, provisionally useful for the science of nature, they quickly became for physics what harnesses become for children: fetters. Had the human spirit not broken those fetters, it would not have been able to surpass Aristotle in physics and Ptolemy in astronomy.

Now, what has broken these fetters? Christianity. Who, above all, profited by the liberty so acquired for pushing on for the discovery of a new science? The scholastics. Who, in the middle of the 14th century, dared to declare that the heavens were not at all moved by divine or angelic intelligences but by an indestructible impulsion received from God at the moment of creation? A Master of Arts in Paris, John Buridan. Who in 1377 has declared the diurnal motion of the earth, a motion more simple and satisfactory for the mind than the diurnal motion of the heavens? Who has neatly refuted all the objections raised against the former of these movements? another Master in Paris, later the bishop of Lisieux, Nicole Oresme. Who has founded the dynamics, discovered the law of the fall of bodies, posed the foundations of geology? The Parisian scholastic and in times when the Catholic orthodoxy of the Sorbonne was proverbial all over the world. Which role was played in the formation of modern science by those free, much vaunted minds of the Renaissance? In their superstitious and routine admiration of Antiquity, they ignored and disdained all the fertile ideas formulated by the scholasticism of the 14th century, so that they might espouse the least defensible theories of Platonic or peripatetic physics. What was, at the end of the 16th and in the beginning of the 17th century, that grand intellectual movement which produced doctrines admitted from then on? A pure and simple return to the teaching which, during the Middle Ages the Scholastic of Paris, presented, so that Copernicus and Galileo are the continuators and disciples of Nicole Oresme and John Buridan. If therefore that science, of which we are rightly proud, could be born, it was because the Catholic Church served as a midwife.

Such are the refutations which in history as well as in logic we must oppose to deceitful and widespread affirmations. Don't you think, Reverend Father, that this would be one of the most important roles, perhaps the essential role, which the future Institut de Philosophie should play? This is why I think that two Chairs would rightly have their place in that Institute: One devoted to the analysis of the logical methods by which the various sciences make their progress would show us that one can, without contradiction and incoherence, pursue the acquisition of positive [scientific] knowledge and, at the same time, meditate on religious truths. The [instruction given from the] other Chair would, by following the historical course of the development of human knowledge, lead us to recognize that in times when men were intent above all on the Kingdom of God and His justice, God gave them for good measure the most profound and seminal thoughts concerning things of this world.

Do you consider me audacious for having thus communicated to you my wishes? Certainly not; because you know that the sole concern that guides me in this matter is the desire to see the Kingdom of God established among us. For such a goal there is no audacity that, instead of being merely allowed, would not be obligatory.

At any rate, as at this moment, with an eye on the intellectual anarchy in which the human mind wallows, I cry to God, *Adveniat regnum tuum*, I seem to hear your prayer as if echoing mine. Let us be heard! This is the wish I make in offering my most respectful regards.

                                                        Pierre Duhem.

16

Countless Steps in a Discovery

Introduction to A. Maire, *L'oeuvre scientifique de Blaise Pascal: Bibliographie critique et analyse de tous les travaux qui s'y rapportent* (Paris. A. Hermann, 1912), pp. i–ix (**1912–10**).

Between a scientific discovery and the one who made it the connection is fairly weak. In many cases time quickly dissolves the tie. Sometimes the treatises and textbooks keep for centuries adding to a mathematical proposition the name of its discoverer or to a law of physics the name of the one who first enunciated it. Thus one says: the theorems of Apollonius, the principle of Huygens; but about the one who carried that name who inquires apart from some curious erudites? At what time in what place did he live? Who was he? By what series of meditations and tries did he arrive at knowing that truth for which he does not become entirely forgotten? These are questions that one hardly thinks of posing, questions that do not make one suffer if they remain unanswered. The proper name that one glues to a proposition is but a convenient tag to designate it. The geometer says Pythagoras' theorem, Simson's theorem, just as he says the theorem of three perpendiculars.

The one who uses a theorem of geometry or a principle of physics is not concerned at all whether it is truly the work of the author whose name it carries. If he were to be concerned, how many unjustified appellations would be revealed to him! Here, one would learn that a truth is attributed to such and such discoverer although he merely suspected it and that his successors had almost entirely discovered it.

There, on the contrary, he would see the true inventor to be entirely forgotten while the idea of which he is robbed assures renown to some popularizer of no merit, to some unscrupulous plagiarizer. The continent discovered by Christopher Columbus receives the name of Amerigo Vespucci. At any rate, apart from these truths poorly named, he would find lots of others and no less significant that are not marked by any name. Nobody demands any longer who are the men whose labor gave science those propositions.

From the part of those who make use of the invention, this indifference with respect to the inventor may seem strange when compared with the meticulous care with which the particulars of the life of great artists and famous literary figures are investigated. Why, here, this insatiable curiosity, carried at times to the point of indiscretion? Why, there, this indifference which borders on ingratitude?

One will perhaps invoke as explanation the essentially objective value of scientific discoveries. In order to apply with the greatest precision the principle of Archimedes, to direct with the highest competence the construction of a ship, the engineeer has no need to know the anecdote about the crown of Gelon, tyrant of Syracuse.

This explanation would be unsatisfactory. Many works of art or of literature possess an intrinsic value and in order to be ravished by it we need not know the history of the author. One can admire the *Iliad* and gaze at the Venus of Milo, although the very existence of the old blind poet can be called in doubt and the name of the sculptor is altogether unknown. Ignorance where we are with respect to the author does not prevent us from enjoying the masterpiece. Still that ignorance does not turn into indifference. How many efforts had been made to find out something about the life of Homer ever since the time when seven Greek cities claimed the honor of having been his birthplace. And if tomorrow some archeologist deciphered in an ancient inscription the name of the one who sculpted the Venus of Milo with what enthusiasm that finding would be celebrated!

Why, then, when it comes to a masterpiece of poetry or sculpture, are we faced with those desperate efforts to discover the name, the century, the country of the author? Why, on the contrary, when it comes to a scientific truth, this carelessness about history which often a little research would easily allow to reconstruct? There must be some reason for such a disparity.

Here is, we think, the reason. While the masterpiece of art or of literature is essentially a personal production and, so to speak, the creation of the author, the scientific discovery is, very often, a collective work.

The mathematician or the physicist hesitates to name the author of a truth in mathematics, because he confusedly feels that by naming him he will commit a sort of injustice which shortchanges all those who directly or indirectly have collaborated with the discoverer. Very often these collaborators are legion.

Thus there comes first the long series of forerunners.

No scientific discovery is a creation *ex nihilo*. It is essentially a composition and combination of elements which had already existed but which now reorganize according to a new plan. Among these elements there are some which have been known for a very long time, so that one has to ascend very far in the past to see their very first generation. Then a gradual descent lets us follow their slow transformations to the point where, fully developed, these elements are ready to rejoin, unite, and constitute a new doctrine. Numerous are those who step by step have prepared that doctrine. They had no knowledge of it, they did not even foresee it except in a confused manner. Still, without their efforts it could not have been born.

When these forerunners, by a labor perhaps centuries long, have cut the materials of the edifice and brought them to its place, the inventor seems to be assembling them, but that name, the inventor, does not always designate a single person. Sometimes one has to take him in a collective sense. To several people who did not know one another, whom distance and language separated, the same idea could occur almost at the same time, so that it is not possible to attribute priority to one over the others.

Even when the discovery of a new truth is the work of single person, that truth does not for long remain a private property. No sooner has the inventor made it public than it becomes a public domain which everybody has the right to cultivate. Numerous workers present themselves who turn up the soil of that land in every sense, who devote themselves to the discussion of the newly announced proposition. Some try to develop it, to expand it, to draw from it many consequences. Others criticize it, denigrate it, pretend to have established its falsity. These contribute no less than do the others to the progress of the new idea because it is the adversaries of a doctrine who force its defenders to make better, clearer, more certain the idea about which they were the first to be convinced.

For most of the time the inventor takes an active part in these discussions. That part is not, however, essential, it is not always the most important. One can manifest more genius in defending the discovery of another than one did by making the discovery. At any rate, when age and death have reduced the inventor to silence, the discussion goes on for a long time about the proposition which he was the first to formulate.

Finally the time comes when the discussion stops, when the proposition is admitted without contest. Is this the end of the collective work of which the idea has, up to the moment, been the object? Not at all, because very often that work becomes even more active. It is now the task to derive from a principle universally admitted all the consequences, all the applications with which it is pregnant. Among these consequences, among these applications how many are there which the inventor did in no way suspect, although they were implicitly contained in what he had conceived! By that centuries–long labor, the principle develops and transforms to such extent that it can hardly be recognized under its primordial form when one encounters it in the writings of those who first discerned it.

All this is so much a matter of common observation as to make it almost useless to cite examples. Do we, however, want to find one which is particularly to the point? The discovery that proved the possibility of vacuum, provides it. Recent debates refreshed everybody's memory about several aspects of that discovery. But to retrace its entire history, it is necessary to go back far in history.

One should read, first of all, that fourth book of *Physics* where Aristotle discusses the question of vacuum. Against the atomists the Stagerite argues to show that the existence of empty space is a sheer contradiction in logic. All procedure that would end in vacuum is inconceivable and absurd.

Apart from John Philoponus all those who commented on the *Physics* — Greeks, Arabs Jews, Christians — accept that doctrine up to the moment when it is to clash with Catholic theology.

To say that the existence of empty space is a contradiction in logic is to say that God himself cannot realize that existence, that all motion that would have a vacuum for its result exceeds the measure of God's omnipotence.

In 1277 the bishop of Paris, Etienne Tempier, condemns this proposition as erroneous. Immediately in Oxford, as well as in Paris, the Scholastic masters outline a new theory of vacuum. The existence of empty space, of dimensions without body, no longer seems to them to be contradictory. God can realize it. If one never encounters a place wholly void of bodies, it is simply because natural forces prevent it from becoming a reality. Not only all bodies, heavy or light, tend, as Aristotle wanted, to return to their natural places once removed from it, but each body tends to remain contiguous with the body which, once displaced, moves in front of it, so that no empty space remains between the one that precedes and the one that follows. This tendency, more powerful than gravity or levity, opposes these when necessary.

We find this doctrine exposed already in the first half of the 14th century by John Dumbleton. He likens the tendency of a body to remain attached to the one that precedes it to the tendency whereby the iron adheres to the magnet. Others, after John Dumbleton, espouse this view. In the middle of the 16th century Julius César Scaliger develops it with complacency.

But then the mechanics note the impotence in which a pump finds itself when it tries to raise water above 32 feet. Solomon of Caus considers this observation as common knowledge and the makers of wells in Florence consult Galileo about it. The tendency for bodies to remain contiguous, a tendency known as the *horror vacui*, is not therefore always more powerful than gravity. This force can become sufficiently large to overcome it. This is what Galileo does not hesitate to state.

At the time when this idea appears, another idea begins to form. What constrains a liquid to penetrate in spite of its weight, into a space, which would, in the absence of the liquid's ascent, remain empty, is not the tendency to follow the body that escapes. It is rather a pressure exercised, in virtue of gravity, by the exterior fluid. We see this idea turn up about the same time in most diverse thinkers. Jean Rey, Isaac Beckman, Descartes, Gianbattista Baliani conceive it in forms more or less precise.

The work of forerunners is then done. All is ready for the work of the inventor to begin.

It begins by Toricelli's celebrated experiment which Father Valeriano Magni may have also thought of independently.

No sooner had Toricelli communicated his discovery to his friend Michelangelo Ricci, and no sooner had the latter informed about it several Italians and Frenchmen interested in physics, than an extraordinary ferment begins to agitate the Europe of the learned. Everybody wants to see "the experiment of quicksilver." Everybody tries to repeat it, modify it, interpret it. Everybody knows what that ferment was in France among physicists whom their common friendship with Father Mersenne tied together. Everybody knows, thanks to the articles published by F. Mathieu and the polemics they created. Pascal and Pecquet tried to have prevail the explication by the "column of air." Roberval opted for the hypothesis of an attractive force, in line with the old opinion of John Dumbleton. The Jesuits rallied to the defense of the impossibility of void, equally affirmed by peripatetism and Cartesianism, two doctrines that tried to harmonize at that time.

The experiment of Puy de Dome put an end to that debate. The merit of having prepared it and of assuring its execution goes to Pascal and his brother-in-law, Périer. But Mersenne, Descartes, and Pascal can dispute who had thought of it first.

The basic propositions of the theory of vacuum and of atmospheric pressure are henceforth secured. It now remains to produce their consequences.

They will lead, first of all, to the invention of the pneumatic machine. Robert Boyle puts together his apparatus to produce vacuum early enough for Pascal to celebrate that discovery. But Otto Guericke goes public with his discovery only after he had pondered it a long time and made use of it in order to carry out his surprising experiments.

To the art of making a vacuum we see quickly joined the art of rarifying and condensing gases through changes of pressure or of temperature. Thus an entirely new branch of experimental physics rises from the work of Pascal. Boyle and his disciple Townley have determined the first shoot of this new branch. Mariotte, Parent, Amontons carried on with its development which, one day, shall lead to Gay-Lussac and Victor Regnault.

It is not enough, at any rate, to view all this from the side of experimental physics in order to see all the fruits to which Pascal's ideas gave birth. Its seeds did not prove less fertile in the field of theoretical physics. In order to explain, with the help of atmospheric pressure, the experiment of quicksilver, it was necessary to specify the notion of pressure, and to clarify and coordinate the principles of hydrostatics. Such was the objective of the *Traité de l'équilibre des liqueurs*, an indispensable preface to the *Traité de la pesanteur de la masse de l'air*. But one would judge altogether wrongly the novelty and importance of this *Traité de l'équilibre des liqueurs*, if one studied it in isolation, by detaching it, as a ring which one breaks off from the chain of tradition that preceded and followed it. How could one evaluate the manner and the extent of the originality of that treatise if one failed to investigate what it received from the very distant Leonardo da Vinci through the mediation of Gianbattista Benedetti, of Benedetto Castelli, of Galileo, of Torricelli, and of Mersenne, if one failed to note at all what it borrowed from Simon Stevin? How, on the other hand, could one evaluate the bearing of the principles set forth in that treatise if one failed to follow them up to the point where they yielded their consequences, where they produced the hydrostatics of Bouguer and Clairaut, of Euler, of Lagrange, and of Laplace?

Who would dare then take such and such proposition of such and such treatise and affirm, with a peremptory assurance, that this truth is from Pascal but that truth is not from him but from someone else? Which sight would be sufficiently penetrating and certain to separate what Pascal let sally forth from his marvelous genius and what his reading of Galileo and Stevin, his conversations with Father Mersenne, the objections of Descartes or Roberval suggested to him? Truly, the discovery of each mathematical theorem, of each principle of physics appears to us as a work to which so many different people have contributed that, by assigning a unique and distinct inventor to this theorem or that principle, we would do multiple injustice.

If the scientific invention is not at all the spontaneous sprouting of an isolated and autonomous genius, if it is a collective and, so to speak, a social work, we have to explore a particularly vast domain each time when we want to retrace the history of a discovery. It will not suffice, far from it, to meditate on the writings of the one to whom common opinion attributes that discovery. We must investigate, read, compare the books of all those who, more or less directly, have been the auxiliaries of that person. We must do the same in respect to the forerunners who had prepared a new idea; to the collaborators who seconded the inventor; to the opponents who forced him to specify, clarify, and reassert his idea; to the successors who put in evidence the latent fecundity of that idea. We have to survey those of whom our author has spoken, those with whom he spoke, those who spoke of him. To make an intelligent reading of a small book such as the *Traité de l'équilibre des liqueurs* or the *Traité de la pesanteur de la masse* will oblige us to explore entire libraries.

Then we shall ardently wish that a sure guide direct us in that labor, that he mark us the page where, concisely or obscurely, lies a phrase pregnant of discovery, that he tell us the volume that contains, in variance with that discovery, an objection, once momentous but now irrelevant; that he names for us the libraries where essential and extremely rare items are conserved. All too often in vain we call for such a guide. To play that role that guide has to possess, at the same time, a vast competence and an indefatigable self-denial that are very rarely found together.

Those who desire, in the future, to follow the steps whereby the genius of Pascal has arrived at his scientific discoveries, those who want to retrace the turns whereby those discoveries reached the point where they became generally accepted, they will not experience a similar anxiety. Thanks to Mr Maire, they will know exactly which books they should read; they will know where those books can be found. Someone, who often and for a long time had to wander without a guide across the unexplored solitudes of the history of science, wanted to remind them of the debt they have incurred toward the author of the bibliography of Pascal.

Cabrespine, August 9, 1911   P. Duhem

17

A Physicist to a Metaphysician

Letter of Duhem of April 11, 1913, to the Rev. R. Garrigou-Lagrange. First
published in the latter's *Dieu. Son existence et sa nature* (Paris: Beauchesne,
1914), pp. 761-63 (**1914-11**). Quoted from the English translation, *God: His
Existence and His Nature* (St. Louis: Herder, 1936), vol. 1, pp. 449-451.

In order to help the reader grasp the exact meaning of Duhem's letter
reproduced here below, it may be useful to preface it with excerpts from
Garrigou-Lagrange's own introduction:

"The principle of the conservation of energy is expressed as follows:
'In a system of bodies removed from all external influence, the total
energy (actual and potential) of this system remains constant.' This
principle is . . . tantamount to saying that it is impossible for motion ever
to cease; if it disappears under one form it reappears under another; thus
the motion of a projectile ceases only in generating heat, and heat itself
produces local motion. The equivalence is established by reason of the
corrective administered to it by the law of the diminution of energy.

Does it follow that a given snap of the finger made a thousand years
ago has still its effect today because of the transformations of energy, and
that it will always be so, without any need for the energy to be *renewed*?
Is it enough to admit that this energy is *conserved* by God, as Descartes
says, and that the *divine motion is not necessary* for the perpetual transformation of energy? . . . To maintain with Descartes that for this, it is
sufficient that God *conserve* the motion, we must understand by this
expression that God continues to move [things].

Thus only can the mechanical principles of inertia and conservation
of energy be reconciled with the metaphysical principle of causality.
Every other reconciliation, which rejects the necessity of the intervention
of the first cause, is illusive.

It is not for the physicist to solve this problem; he cannot pronounce finally on the validity of the solution given by the metaphysics of the Schools; he must merely recognize that this solution is in no way opposed to what physics has the right to affirm about the validity of its principles in the phenomenal order.

On this point it gives us pleasure to publish a letter from Pierre Duhem, of the Academy of Sciences, in which he gives us a summary of the main ideas of his fine work, *La théorie physique*.

The letter reads as follows:"

Dear Father: I owe you some explanation for certain ambiguous terms in my previous letter and especially for the name '*axiom*' or '*so-called axiom*' which I gave to the principle of inertia.

I begin by stating precisely that I shall take the words mathematics, physics, and metaphysics according to the meaning generally given them by our contemporaries, not according to the meaning given them by Aristotle and the Scholastics.

In these circumstances, the law of inertia does not exist for the mathematician; the principles of the science of numbers and of geometry are the only ones that he has to admit; he is not concerned with the principles of mechanics and physics; if he happens to study the problems presented to him by the mechanist and the physicist, he does so regardless of the way by which they have been led to formulate these problems.

I consider, therefore, the principle of inertia only as it is for the physicist.

One may say of it, then, what may be said of all principles of the mechanical and physical theories. These fundamental principles or *hypotheses* (in the etymological sense of the word) are not *axioms*, self-evident truths. Nor are they *laws*, that is, general propositions reached directly by induction from the teachings of experience.

It may be that certain rational probabilities or certain facts of experience *suggest* them to us; but this suggestion is in no way a *demonstration*; it does not confer on them, of itself, any certitude. *From the point of view of pure logic*, the fundamental principles of the theories of mechanics and physics can be looked upon only as *postulates freely posited by the mind.*

From the ensemble of these postulates, deductive reasoning deduces an ensemble of more or less remote consequences which agree with the perceived phenomena; *this agreement is all that the physicist expects* from his *postulated principles.*

This agreement confers a certain probability upon the fundamental principles of the theory. But it can never confer certitude on them, for it can never be demonstrated that, if other postulates were taken as principles, consequences would not be deduced which would agree just as well with the facts.

Besides, it can never be affirmed that some day new facts will not be discovered which no longer agree with the consequences of the postulates that had been posited as being at the basis of the theory: new facts compelling us to deduce a new theory from new postulates. This change of postulates has been effected many a time in the course of the development of science.

From these considerations two consequences follow: (1) *We shall never have the right to affirm categorically of any one of the principles of the mechanical and physical theory, that it is true.* (2) *We are not allowed to affirm of any one of the principles on which the mechanical and physical theory rests, that it is false, so long as there has been no discovery of phenomena that disagree with the consequences of the deduction of which this principle constitutes one of the premises.*

What I have just said applies particularly to the principle of inertia. *The physicist has not the right to say it is certainly true*; but still less has he the right to say it is false, since we have so far met with no phenomenon (if we leave out of consideration the circumstances in which the *free will* of man intervenes) that compels us to construe a physical theory from which this principle would be excluded.

All this is said without going beyond the domain of the *physicist,* for whom the principles are not affirmations of real properties of the bodies, but premises of deductions the consequences of which must be in agreement with the phenomena every time that a free will does not intervene to dissarange the determinism of the latter.

To these principles of physics, can we and must we make certain propositions correspond which would affirm certain real properties of bodies? To the law of inertia, for instance, must we make the affirmation correspond that there is, in every body in motion, a certain reality, an *impetus*, endowed with such or such characteristics? Do these propositions apply or not to other beings endowed with free will? These are problems that the method of the physicist is incapable of grappling with and it leaves them to the free discussion of the metaphysicians.

There is only one case which would induce the physicist to be opposed to this liberty of the metaphysician. It is that in which the metaphysician would formulate a proposition directly contradicting the phenomena or a proposition which, introduced in virtue of a principle in the physical theory, would lead to consequences in contradiction to the phenomena. In this case, there would be just grounds for denying the metaphysician the right to formulate such a proposition.

Now you have, Reverend Father, the summary of what I would say if I were ever to write, concerning the principle of inertia, the article that you so kindly wish me to write.

P. Duhem.

18

The Slow Evolution of Science

Le Système du monde. Histoire des doctrines cosmologiques de Platon à Copernic. Tome I. *La cosmologie hellénique* (Paris: A. Hermann et Fils, 1913), pp. 5-6 **(1913-2).**

In the genesis of a scientific doctrine there is no absolute starting point. No matter how far into the past is traced the line of ideas that prepared, suggested, and foreshadowed that doctrine, one always reaches opinions that, in their turn, have been prepared, suggested, and foreshadowed. And if one stops following this chain of ideas that preceded one another, it is not because one has on hand the initial link, but because the chain disappears by sinking into the depths of an unfathomable past.

All astronomy of the Middle Ages has contributed to the formation of the system of Copernicus. By the mediation of Islamic science, the astronomy of the Middle Ages is tied to Hellenic doctrines. The most elaborate Hellenic doctrines, those that are best known to us, derive from the teachings of ancient schools of which we know very little. These schools in their turn, had inherited the astronomical theories of the Egyptians, Assyrians, Chaldeans, Indians — theories of which we know almost nothing. The night of past centuries is almost on hand and we still feel to be very far from the first men who observed the course of stars, registered their regularities, and tried to formulate the rules which they obey.

Unable to reach far back to a very first principle, we are forced to select an arbitrary starting point to the history which we want to retrace.

We therefore will not ask what were the astronomical hypotheses of the very ancient peoples — Egyptians, Indians, Chaldeans, Assyrians. The documents where their hypotheses are set forth are few and far between. The interpretation given of them is all too often so arbitrary as to make sceptical even the most learned. At any rate, we lack the competence not only to evaluate but even to summarize the discussions of orientalists and egyptologists.

We will not report either, except in generalities, what can be reconstructed from the doctrines of the ancient sages of Greece. The short fragments, at times of doubtful authenticity, to which their works have been reduced, hardly allow us to guess how their ideas were born one after another, and how each of them developed.

Decidedly, it is with Plato that we shall begin this history of cosmological hypotheses. He is the first philosopher whose writings, useful for our purpose, have come down in an integeral and authentic form. He is, therefore, the first whose thought about celestial motions we can know in entirety, or at least all that he wanted us to know about it.

But, right away, we also see what is arbitrary, and even partly rational, in the choice of such a point of departure. In order to understand Plato's astronomical theories, it is not enough to study Plato because these theories did not generate themselves. They take their starting points elsewhere and derive from earlier sources. What Plato teaches about celestial motions is constantly inspired by the teaching of the Pythagorean schools and in order to understand well the astronomy of the Academy, it is necessary to become first familiar with the astronomy of Magna Graecia.

Thus are we lead to say a few words about the astronomical doctrines that were received among the Pythagoreans so that we may better understand those to be taught by Plato.

19

The Great Year Destroyed by the Church

Le système du monde. . . Tome II. La cosmologie hellénique (1914), pp. 390, 407–408 **(1914–1)**.

In the system which Maimonides sets forth we see, so to speak, the culmination of all the ideas whose development has been traced in this chapter.

We find there, first of all, the affirmation of the principle that Aristotle had already formulated with such clarity: The various parts of the universe are interconnected by a rigorous determinism and this determinism subjects the entire world of generation and corruption to the rule of celestial circulations.

We find there the corollary of that principle, namely, the definition of an astrological science which ties all changes accomplished here below to the motion of a specific planet.

We see there the preponderant role which that astrology attributes to the Moon as a rule of water and humid matter. The Moon forces them to grow and decrease with her. The theory of tides clearly proves the reality of this lunar action and, through it, of all influence emanating from the celestial bodies.

Finally, we hear stated that the very slow changes on earth are tied to the almost imperceptibly slow motion of the fixed stars whose revolution measures the Great Year.

To that system all the disciples of Greek philosophy — Peripatetics, Stoics, Neoplatonists — have contributed. To that system Abu Masar offered the homage of the Arabs. The most illustrious rabbis, from Philo of Alexandria to Maimonides accepted that system.

Christianity was needed to condemn that system as a monstruous superstition and to throw it overboard. . . .

Hardly anxious to explore in detail the works of Greek astronomers, the bishop of Hippo and with him, undoubtedly, the great majority of the Church Fathers, did not know how to separate, in a precise manner, the hypotheses of the astronomers from the astrologers's superstitions. The former were confusedly included in the disapprovals accorded to the latter. . . .

Let us not therefore search in the writings of the Church Fathers for the traces of a meticulously and sophisticatedly treated science. We assuredly cannot find them there at all.

Let us not, however, neglect the little they said about physics and astronomy.

First of all, their teachings on this topic are the first seeds from which the cosmology of the Christian Middle Ages would slowly and gradually develop.

Also, and above all, the Church Fathers hit, and did so in the name of the Christian Creed, the pagan philosophers on points which, today, we consider more metaphysical than physical but where actually lie the cornerstones of the physics of Antiquity: such are the theory of an eternal prime matter, the belief in the stars' domination over sublunary things and in the periodic life of a cosmos subject to the rhythm of the Great Year. By destroying through these attacks the cosmologies of peripatetism, of Stoicism, and of Neoplatonism, the Fathers of the Church clearly prepare the way for modern science.

20

Science and the Good Shepherd

Le système du monde . . . Tome IV. L'astronomie Latine au Moyen Age. La crue de l'Aristotélisme, pp. 452–453 (**1916–1**).

Apart from their prime matter, which they get from the active Intellect, the sublunary things receive, therefore, from the heavens and from their motors all the goods they have right to. "There are men," states Avicenna, "who pretend that the diversity of celestial configurations and motions seems to be produced with the intent of generating the corruptible things." Avicenna then develops the motives invoked in favor of this opinion: The heavens move so that they may resemble as closely as possible the supreme Cause. Now the supreme Cause is the source from which all good things in this world originate. This is why the heavens want to be similar to It, so that what moves them is the desire to spread those good and fertile influences within the sublunary matter. In this view we find the trace of the teaching which *The Theology of Aristotle* borrowed from Christianity. It affirms in a particular case this desire which every being experiences to produce, in an inferior being, the good of which it is capable.

Avicenna and al-Ghazzali, who combatted in a general way the teaching of *The Theology of Aristotle*, rise just as heatedly against this particular consequence of it.

"It is impossible," states Avicenna, "that the intent of creating bodies susceptible of generation and corruption be the cause of the existence of celestial bodies and of their motions around these inferior beings. Whoever sets out in search of another view is assuredly a viler being than the object of his inquiry. For it would then result that the celestial bodies are viler than the inferior substances. . . . How could it be the Sun's intention to produce these vile beings, and how these vile beings could be the goal toward which tends its eternal motion? . . . Never does what is noble tend toward what is vile for the sake of that vile thing. But, then, if anyone desirous of another view is more vile than that very view, is then a shepherd too viler than his sheep, a teacher viler than his disciples, and the prophet viler than his people? Because the shepherd cares only for his sheep, the teacher only for his disciples, and the prophet for his people. We answer by noting that the shepherd, inasmuch as he is a shepherd, is viler than his sheep, but more noble insofar as he is a man. . . . If one considers in him only what makes him a shepherd, he is assuredly viler than his sheep. And one can say the same of the teacher and the prophet too. The nobleness of the prophet consists in that which makes him perfect in himself. It is in terms of his qualities that he is noble and not because he teaches the people. If one considers in him only that he teaches the people, it would follow that those whom he seeks to instruct are more noble than the one who instructs them."

Assuredly, no philosopher, outside the influence of Christianity, could make intelligible the benevolence whereby the superior being, with no detriment to it, can desire the good of the inferior being. Not one of those philosophers could comprehend that the prophet loves his people, that the Good Shepherd loves His sheep to the point of giving His life for them.

21

Christian Common Sense.

Le système du monde. . . . Tome V. La crue de l'Aristotélisme, p. 580 (**1917–1**).

Siger of Brabant did not pretend that he had taught the truth to his contemporaries, but only that he had set forth for them the thought of Aristotle. Did he achieve his aim? It would be childish to contest him. Even when his propositions could not be authorized by a specific text of Aristotle, they were very clearly in the sense of Peripatetic philosophy. Siger's reasonings are dominated by this idea: The possibility of existing is already in a certain way to exist in potency, and potential existence is what matter is. Now this is the guiding principle of all of Aristotle's philosophy, the very principle which one cannot admit without becoming to some extent a peripatetic, nor can one reject it without ceasing to be a peripatetic. Let us speak frankly: of all authors of the 13th century, Siger of Brabant is the only one who presents us the authentic thought of Aristotle, without mixture, without deformation, without reticence.

In the writings of Albert the Great, and even more so in the writings of Saint Thomas Aquinas, Peripatetism appears in an attenuated form as being impregnated with Neoplatonism and divested of several of its essential theses. Then it seemed possible to reconcile it with the teaching of the Catholic Church. Those who like Siger of Brabant enrolled themselves in the school of

Averroes, simply reinstated a different peripatetism. They restored to Aristotle's system its rigidity of logic.

Faced with an Aristotelianism, which now showed itself without disguise and completely naked, Christian scholasticism understood what were to be its conditions as a victor if it let itself be conquered by that doctrine. Either it would have had to declare as false all the teachings of the Catholic Church, or it would have had to admit that two contradictory teachings could be equally true: one, because the Church teaches it; the other, because demonstrated by the reasonings of the philosophers. It would have had, in a word, to sacrifice its faith or its common sense. Its choice was quickly made: it sacrificed pagan philosophy.

22

The Creator's Liberty
and the Liberation of Science

Le système du monde. . . . Tome VI. Le reflux de l'Aristotélisme, p. 66 (**1954-2**).

These diverse condemnations touched on very formal and essential doctrines of Aristotelianism and Averroism, doctrines whereby astrology could claim to be part of those philosophies. Two other condemnations hit peripatetic physics at its very center. They destroyed the theses that served so to speak as the foundation of the entire edifice. They in fact excommunicated anyone who adhered to these two propositions: "God could not impart to the heavens a [linear] translational motion because the heavens, thus moved, would have left a vacuum behind them," and "The First Cause could not have created several worlds."

By hitting these propositions with an anathema, Etienne Tempier and his council declared that in order to remain subject to the teaching of the Church and not to impose limits to God's omnipotence, one had to reject peripatetic physics. By that move they implicitly called for the creation of a new physics which the reason of Christians could accept. We shall see the University of Paris try to construct this new physics during the 14th century; by its efforts it laid the foundations of modern science. The latter was born, so to speak, on March 7, 1277, from the decree issued by Monseigneur Etienne, bishop of Paris. One of the principal objectives of the present work would be to justify this assertion.

23

Medieval Positivism

Le système du monde. . . . Tome VI. Le reflux de l'Aristotélisme, pp. 728–729 (1954–2).

In 1511, or the times when Lionel Coronel published his book, the Masters of Arts were no longer held by oath to avoid theological discussions. They had been held by such oath in the time of Buridan and the teachers of theology reminded them of this whenever necessary. One should not therefore be surprised that the philosopher from Béthune does not tell us about his reasons for believing in the teaching of the Church in the manner in which he set forth his reasons to stick with the principles of physics.

At any rate, Buridan let us know with admirable clarity the nature and degree of confidence which it is appropriate to grant to the principles of physics. Those principles, even the most important and most general of them, even the one that decomposes all substance into matter and form, are not necessary truths that one could deny without getting involved in absurdity. Such are the propositions drawn from experience. One has recognized them to be exactly valid in a great number of particular cases. One has not found circumstances in which they were contradicted. Then the intellect, pushed by its natural tendency toward the true, has proclaimed their universality.

With great clarity, with great precision, Buridan has described us his philosophical method. That method admits to be incapable of giving, in metaphysics, demonstrations that reach their conclusions in an irrefutable manner. That method will therefore humbly bow before the teachings of Faith which alone are in a position to give to essential questions secure answers. That method recognizes to be powerless to discover, in physics, in an *a priori* manner the effects we observe. That method will, therefore, limit itself to procede *a posteriori*, to establish by induction the laws of experimental origin, to combine hypotheses with which we are satisfied when they have saved, in the simplest possible manner, all the phenomena. After many vicissitudes, Christian Faith and experimental Science have triumphed over Aristotelian dogmatism as well as Ockhamist Pyrrhonism. Their combined efforts generated that Christian positivism whose rules Buridan made known to us.

This positivism will not be practiced by Buridan alone. It will be espoused by his disciples, by Albert of Saxony, Temon, son of the Jew, by Nicole Oresme, by Marsilius of Inghen. These are the men who will create the Parisian physics, this first sketch of modern science, and it is by that method that they will create it.

24

From *impetus* to Newton

*Le Système du monde. . . . Tome VII. La physique parisienne au XIV*ᵉ *siècle,* pp. 298-299 (**1956-1**).

Let us close the *Lectiones* of Saint Thomas Aquinas and let us open the *Quaestiones* of Albert of Saxony. What do we read there?

It is not the disturbed air that maintains the movement of the projectile. The air which the projectile has to divide in its course has no moving force. It merely plays the role of resistance. What moves the projectile is the *impetus* which the throwing force has communicated to that body. The greater is this *impetus*, the greater is the velocity. For different bodies with equal velocities, the *impetus* is proportional to the mass of the body.

The violent [forced] motion does not last indefinitely because the *impetus*, by struggling against the natural gravity of the projectile and against the resistance of the medium, keeps weakening.

It is altogether otherwise in natural motion. There the *impetus* keeps increasing and, consequently, so does the velocity of the thing in motion. It is this reason, and not the continual increase of gravity, which keeps accelerating the fall of a heavy body. The mathematical formula of that acceleration is already surmised.

The alternatives of the increase and decrease of the *impetus* in a motion which, alternately, is natural and forced, accounts for the oscillations of a body around its position of equilibrium.

Such is, in a few words, the balance of the acquisitions made by the science of mechanics of the Parisians, between the time of Saint Thomas Aquinas and the time of Albert of Saxony. The dynamics of Aristotle has been been completely overturned. The foundations have been laid for a dynamics which will be that of Galileo, of Descartes, of Pierre Gassendi, of Torricelli, a dynamics waiting to become the dynamics of Huygens, of Leibniz, and of Newton.

25

The Church and Freedom of Thought

Le Système du monde. . . . *Tome X. La cosmologie du XVe siècle. Ecoles et universités au XVe siècle*, pp. 323–324 (**1959-1**).

Of the diverse elements that compose a star, such as the earth, some are heavy and tend towards a certain point, while some are light and flee that point. The entire star neither approaches that point nor moves away from it. The star is neither heavy nor light because the gravity of some of its elements is exactly compensated by the lightness of the others. Thanks to the compensation the star remains suspended in space. In order to create the world, God made recourse to the four mathematical sciences — geometry, arithmetic, astronomy, and music. The exact balance of which we talk was the specific work of geometry. Such is, so it seems, the sense one has to attribute to certain passages of Nicolas of Cusa.
. . .

Nicole Oresme admitted that space could contain several systems of which each was composed of an earth surrounded with water, air, and fire. To these diverse words he applied a theory of gravity altogether similiar to the one which Plutarch had proposed. He did not go as far as assigning to the worlds denizens similar to the ones on our earth. As far as is known, nobody, during the Middle Ages, had submitted such a supposition until Nicolas of Cusa, whose imagination did not seem to have limits, came to propose it.

The earth, declared the future bishop of Brixen, is not the vilest among the celestial bodies, and the men, animals, and plants that inhabit it are not inferior in nobility to those who inhabit the Sun and other planets.

Our philosopher recognized, at any rate, that we cannot know much about the beings that inhabit other planets. "We suspect," he says, "that the inhabitants of the Sun are more sunlike, more enlightened, illuminated, and intellectual. We suppose them to be more spiritual than those that are on the Moon who are more lunelike. Finally, on the earth, they are more material and gross. Thus the intellectual beings that are on the Sun are more in act and less in potency, whereas the inhabitants of the earth are more in potency and less in act. As to the Moon's inhabitants, they float between these two extremes."

"These opinions are suggested to us in view of the influence of the Sun which is of igneous nature, by the influence of the Moon which is both aqueous and aerial, and by the earth's more material heaviness."

When for the first time, in Western Christianity, one hears speaking of the plurality of inhabited worlds, one sees the idea proposed by a theologian who had, a few years earlier, taken the floor in an ecumenical Council. He, who in a book soon to become famous, tried to guess the characters of the inhabitants of the Sun and of the Moon, was to be honored by the confidence of popes who succeeded one another in the chair of Peter. The highest ecclesiastical dignities were reserved for him.

Can one wish a more manifest proof of the extreme liberty which the Catholic Church, on the decline of the Middle Ages, left for the meditations of philosophers and for the tentatives of the physicists?

26

In Defense of the French Mind

La Science allemande (Paris: Hermann, 1915), pp. 3-4 and 141-143 (**1915-2**) (translation of John Lyon).

If ever the word "conspire" could be used in the fullness of its sense, it is assuredly of France which quickens under our eyes. Every breast breathes in unison, every heart throbs to the same feelings. One single soul animates this vast body of France. To save and redeem the soil of France, dear students, your elders, your fellow students, have soaked it with a blood that knew no price. A short time ago, I shook the hands of those of you who belong to the class of 1915. And when I said to them, "Farewell! May God protect you!" I saw their eyes glisten with a flash of joy. A young Frenchman is not completely happy in fulfilling his task unless it is very dangerous. And you, the younger fellows of these students, I sometimes see you clench your fists, for you dream of an avenging arm, and you believe that you already possess it. All about you — mothers, wives, sisters, daughters of soldiers — work likewise to alleviate the afflictions of the combatants or the sufferings of the wounded. And if it be that brows must furrow in mourning, these brows appear to us to radiate through their sorrow, for the acceptance of sacrifice casts its halo there.

In the midst of this common aspiration, he who is about to speak to you feels profound anguish. Except through prayer he seems incapable of collaborating in the great public work. M.

l'Abbé Bergereau had compassion on this sorrow, brought about by the sense of uselessness. He said to me: It isn't the soil of France alone that has been invaded. Foreign thought has taken French thought captive. Go sound the charge which will deliver the soul of the Fatherland!

Assigned my combat position, I come running. The position is without danger, and will therefore be without glory. I shall have no occasion to pour out my blood there, but I shall pour out all the devotion that my heart contains.

I come before you to take my humble part in the national defence. . . .

Is it possible to give a conclusion to these various reflections? It issues from what precedes so naturally, it appears, that we feel somewhat humble in formulating it. Therefore we shall do it with extreme brevity.

French science, German science, both deviate from ideal and perfect science, but they deviate in two opposite ways. The one possesses excessively that with which the other is meagerly provided. Here the geometric spirit reduces the subtle intellect to the point of suffocation. There the subtle intellect dispenses too willingly with the geometric spirit.

In order, therefore, that human science might develop in its fullness and subsist in harmonious equilibrium, it is good that one see French science and German science flourishing side by side without trying to supplant each other. Each of them ought to understand that it finds in the other its indispensable complement.

Therefore the French will always find profit in pondering the works of German scholars. They will come across there at times the solid proof of truths which they have discovered and formulated before being properly assured of it, at other times the refutation of errors which an imprudent intuition has caused them to receive.

It will always be of use to Germans to study the writings of French inventors. They will find there, so to speak, the statements of problems to the resolution of which their patient analysis ought to apply itself. They will hear there the protestations of common sense against the excess of their geometric spirit.

That German science in the nineteenth century took its departure from the work of great French thinkers no one from the other side of the Rhine, would, I think, dare to contest. And no one from this side dreams of failing to recognize the contributions with which, later on, this German science has enriched our mathematics, physics, chemistry, and history.

These two sciences, then, ought to respect the harmonious relations that exist between them. It does not follow that they are of the same rank. Intuition discovers truths; demonstration comes after and secures them. The geometric mind gives body to the edifice which the subtle intellect conceived in the first place. Between these two inclinations there is a hierarchy analogous to that which ranks the mason in respect to the architect. The mason only does useful work if he conforms it to the architect's plan. The geometric mind does not pursue fruitful deductions if it does not direct them toward the end which the subtle mind has discerned.

On the other side, on the part of science which constructs the deductive method, the geometric spirit can well assure a rigor without reproach. But the rigor of science is not the truth of it. The subtle intellect alone judges whether the principles of the deduction are admissible, whether the consequences of demonstration are conformable to reality. In order for science to be true, it is not sufficient that it be rigorous; it must depart from good sense in order to return to good sense.

The geometric spirit which inspires German science confers on it the force of a perfect discipline. But this narrowly disciplined method can only come to disastrous results if it continues to put itself under the command of an arbitrary and senseless algebraic imperialism. The instruction which it ought to obey and to receive, if it wishes to do useful and beautiful work is from that which is the principal depository of good sense in the world — that is, from French science. *Scientia germanica ancilla scientiae gallicae.*

27

In Praise of the Feminine

"Discours de M. Duhem." *Groupe Catholique des Etudiantes de l'Université de Bordeaux. Année 1915-1916. Compte rendu de l'Assemblée Générale du 25 Juin 1916* (Bordeaux: Imprimerie Nouvelle F. Pech), pp. 11-18 (**1916-13**).

Mesdemoiselles! A few days ago your President and some of your colleagues, who are particularly dear to me because they are my students, came and asked me to preside at your General Assembly. This request does great honor to me and is, at the same time, the cause for great joy. It indeed allows me to repeat in public what I have said many times in private, namely, to reaffirm the deep sympathy which I feel for your Association, and the great hopes which it inspires in me.

However, this honor and this joy are not without some apprehension. The one who says "please, preside" also says "please, speak" and it does not seem to me easy to address you.

A professor can easily penetrate even the heart of his students, your fellow students. He too was a student; the years they now live, he too has lived. It is enough for him to reminisce in order to become quite similar to those among whom he now finds himself. Between them and him the thoughts and sentiments are very common indeed.

It is not the same when it comes to you, Mesdemoiselles. Of course, I shall not astonish you by saying that I have never been a co-ed. Between us there rises that wall behind which the most loving and most confiding daughter always remains a mystery for her father, indeed, a mysterious enigma.

Only one man can read as an open book the soul of a young woman, because the natural perspicacity is doubled in his case by a supernatural insight. The priest can indeed guide you. With what assurance, with what devotion does his sacred character allow him to do that, you know well. You also feel towards him a profound gratitude. By making me, on your behalf, the spokesman for this gratitude toward your chaplain, I am sure I meet more than half way your innermost wishes.

And since today I have the honor of being your representative, let me register, if not pay, another of your debts. One such debt you have contracted toward those who have been young women as you are. In order to find them similar to you, it is enough for them to remember the time when God was not their only father and their only mother. Indeed, how well do they understand you and second-guess you! They offer you hospitality, and with what tact and delicacy they do so! You are 'chez elles' but they wish that you would pronounce the word so dear to every woman: 'chez moi'.

What only a priest can accomplish, I would not undertake to do. I would not be so daring as to offer you advice. More modestly, I want to submit to you a few reflections touching on what is said of you.

Among you, Mesdemoiselles, the desire to learn grows with no letup and the calling to be a teacher shows itself more and more frequently. You come each year in ever larger number to the various departments of the University in order to gather the parcels of truth which we have the mission to distribute. And immediately, among those who are drawn to matters intellectual, there arise preoccupations as well as worries. To do science and spread it has until now been almost exclusively the business of men. If women undertake an ever great share of this task, will not the manner of carrying it out also undergo some change? These young women who crowd into our classrooms, will not they

impose some new form on our virile methods of instruction? And
this new form, what it will be like? Should we look forward to it,
or should we be apprehensive on its account?

It is, so it seems to me, easy to answer these questions. By
coming to undertake part of our task, you will do, Mesdemoiselles, what you always do wherever you go, and it will be a work
of blessing.

Enter in the house which has not been occupied by a young
woman for some time. Open the door of the living room. The
pieces of furniture you find to be of good style, the pieces of art
of good taste. All is in order. The easy chairs are aligned symmetrically, the curtains are evenly spread. However, nothing ravishes
the eye, nothing pleases it. In this morose place one does not wish
to stay long.

Along comes a young woman. What is she going to do? She
slightly displaces a table, moves another piece of furniture, puts a
bouquet of flowers into a vase, with an easy touch adds another
fold to a curtain. All this, to be sure, is small matter, but all is
changed thereby. In the living room, once so somber and austere,
all shines and smiles now. The light invites your eyes to be fixed
on that watercolor which you have not previously noticed, a ray
of light caresses the strong shape of that bronze which an obscure
corner was hiding. With a young woman charm has come into the
house. Those who dwelt in it do not recognize their old lodgings.
It appears to them that a fairy has turned it into a palace.

Into the manor of studies, you let elegance and clarity enter,
through you this antique habitat becomes lovable and charming.

In the evening, after long hours passed in combing equations
or in deciphering a mansucript, I love to re-read a certain
Institutions de physique printed in 1740. Chapter after chapter, all
whose beginnings are embellished with a fine engraving, I follow
the exposition of the first principles of mechanics. I see them
developing and produce, with Newton, the great laws of the
system of the world, I see them lead a Leibniz to one of the
deepest metaphysical doctrines ever conceived. In all this a lucid
discourse flows from one proposition to another. They assemble
as if by themselves, with no pain, no effort. This bundle of beauti-

ful thoughts recalls the bouquet of wildflowers which painters of old placed, with a skillful negligence, between the fingers of women whose portraits they were doing. The author of *Institutions de physique* had, I imagine, the fine hands needed to hold such a bouquet, because her name was Gabrielle-Emilie de Breteuil, marquise du Chastelet.

When I close the *Institutions de physique* of Mme. du Chastelet, I dream at times: How feminine! Even more often, I think: How French!

It seems indeed to me that the French mind is essentially feminine. I cannot speak of it without imagining a very noble and beautiful lady whose smile and composure are yet all grace and simplicity.

You now see what the French intellect does when it penetrates a doctrine.

The positive mind of the English had discerned there innumerable facts, all minutely observed and described. Yet memory admitted its impotency to retain so many riches and reason no longer saw the benefit it could derive from them.

The German method has erected there learned and rigid theories, but in those narrow compartments, symmetrically grouped in terms of complicated and measured rules, reality did not know where to settle. Certain cabinets remained void of all facts, whereas others could not accommodate the data of observation except at the price of efforts and contorsions.

It is now the turn of the French intellect. With a single gesture it lays out in front of her the accumulated facts. With a single regard she distinguishes those that indicate an outstanding characteristic, those that can be taken for examples. She throws light on them, she arranges them in such a way as to support one another. Then, with a touch of scorn she sweeps out the rubbish of superficial documents. Then she takes stock of theories. She demolishes the wings which preoccupation for vain symmetry has constructed. She simplifies the overcomplicated arrangements. Then reality comes into possesion of a harmonious and clear palace where, henceforth, it can live at ease, where everyone can contemplate it at leisure. In the doctrine assimilated in France those who gathered the facts, those who had scaffolded the systems, will hardly recognize their work, because they always doubted that it can become one day so ordered, simple and beautiful.

Sometime, they become jealous of the one who possessed the secret of similar transformations. Are not they the ones who discovered the facts? Are not they the ones who put together those theories? What new was said by France? And France answers smilingly with Pascal: "Don't tell me that I have said nothing new. The arrangement of the material is new. When one plays tennis it is the same ball which is used by both players, but one places it better." These are, the same young woman would repeat, "the same furniture, the same sculptures, the same paintings but I place them better."

The French intellect has for some years notably forgotten this art, so feminine, of putting each thing in its proper place. The piling up of disordered facts, the graceless rigidity of systems no longer repulse her. She takes pride in being in the midst of trash and in the labyrinths of bizarre setups. It is up to you, Mesdemoiselles, to bring her back to her tradition of order, harmony, and clarity. As workers of the intellect, you should keep, in regard to things of the spirit, that charming skill of those modest sisters of yours who earn their living by needlework. You will deliver a lecture, write a book, with that elegance, with that good taste, indeed with that chic with which they set about to drape a coat, to fashion a ribbon tie. Thanks to them, there is no woman in the world who thinks that she is well dressed if her robe or hat does not come from France. Thanks to you, we shall see again that time when no professor on earth thought to teach well unless he took for guide a French treatise.

In the studies you pursue, in the classes you will give, guard with jealous care your sense of being a young woman, a sense so refined, so assured, so measured. You will do a very French work. Will you do but that? Shall I dare to say that at the same time you accomplish a most Christian task? Certainly, I shall dare to say it and without anyone, I think, being astonished. The French spirit and the Christian spirit are so closely united that to sap one is almost to ruin the other, while to revitalize one is, all too often, to revitalize the other.

What are the dangers posed to the Christian soul by the study of science? They are of several kinds but it seems that they can be grouped under two principal heads.

The laborious mind can easily attach itself in an exclusive manner to the study of facts. Immersed in the meticulous analysis of all details, which can be felt and seen or which can be counted and weighed, such a mind lets the myopia of its attention readily grow. Soon it will be incapable of contemplating an idea. It finds ideas too elevated or far removed. Instead of recognizing there the only truth worth seeking, it takes idle dreams for ideas because for it there is no certainty apart from the witness of the senses. It falls and rolls by the degrees of a suspicious positivism toward an abject materialism.

At times the reasoning spirit falls prey to the play of deductions that lack foundations. It finds pleasure in mounting the narrow and shaky scaffolding of theories. It does not investigate whether this fragile construct rests on secure bases. It only wants the construct to be able to rise to the clouds which it takes for the sky. Soon it is seized by a vertigo and yields to the vaguest pantheism and to the most nebulous mysticism.

You know, Mesdemoiselles, these two dangers. You do not have to search for long "in books and among men" to recognize its very numerous and most deplorable victims. In order not to fall into any of those two excesses, what should they not have gone without? Precisely those qualities which makes the intellect similar to French woman.

Could your taste for elegance ever be contented with a collection of observations or of a batch of documents? How could it fail to claim that idea which alone can introduce order and beauty into a pile of facts? Your horror of all that lacks gracefulness, would it not guard you against these brutalities of materialism?

On the other hand, your finesse, more apt to guess the truth by an intuitive and sudden glance than to conquer it through the marshalling of endless syllogisms, will put you on guard with a marvelous clear-sightedness against trusting sweeping systems. At the end of heavy and pedantic deductions, as they present to you strange and dispiriting conclusions, suddenly there will arise from your spiritual good sense the laughter of a young French woman, a laughter stronger than all refutations.

Thus, Mesdemoiselles, against the crude culture of facts, against the sophisticated skullduggery of reasoning, the appropriate qualities of your intellect will safeguard the faith of your pupils even when the subject matter of your teaching seems to have no connection with the Christian truth.

When the Eternal Reason, when the Word of God wanted to manifest Himself to men, He received from a young woman the body that was to make Him visible, the tongue and the lips that allowed Him to be understood.

For a long time our classes have ceased to be the echo of this Word of God. For a long time man has pretended that the vacillating glimmer of his reason suffices to let him discover truth. This pale light could not guard him against strange aberrations.

On the soil of our beloved country your brothers shed their redemptive blood without counting it. Will the price of such a sacrifice not become the reconciliation of French thought and Christian thought? Are we not going to see again what was a living reality during the Middle Ages and in the century of Pascal and Bossuet? In order to secure its progress along the road which is suited for it, would not the human reason march again, as it did before, with eyes fixed on the divine revelation? And is it not for preparing that Christian renaissance of science, for giving voice to the Word of God, that the Holy Spirit puts in the hearts of so many young women the desire to learn and the sacred vocation to teach?

(continued from p. 3)

By the same author

The Absolute beneath the Relative and Other Essays

The Savior of Science
(Wethersfield Institute Lectures, 1987)

Miracles and Physics

God and the Cosmologists
(Farmington Institute Lectures, Oxford, 1988)

The Only Chaos and Other Essays

The Purpose of It All
(Farmington Institute Lectures, Oxford, 1989)

Catholic Essays

Cosmos in Transition: Studies in the History of Cosmology

Olbers Studies

★ ★ ★

Translations with introduction and notes:

The Ash Wednesday Supper (Giordano Bruno)

*Cosmological Letters on the Arrangement
of the World Edifice* (J.-H. Lambert)

Universal Natural History and Theory of the Heavens (I. Kant)

Note on the Author

Stanley L. Jaki, a Hungarian-born Catholic priest of the Benedictine Order, is Distinguished University Professor at Seton Hall University, South Orange, New Jersey. With doctorates in theology and physics, he has for the past thirty years specialized in the history and philosophy of science. The author of thirty books and nearly a hundred articles, he served as Gifford Lecturer at the University of Edinburgh and as Fremantle Lecturer at Balliol College, Oxford. He has lectured at major universities in the United States, Europe, and Australia. He is honorary member of the Pontifical Academy of Sciences, *membre correspondant* of the Académie Nationale des Sciences, Belles-Lettres et Arts of Bordeaux, and the recipient of the Lecomte du Nouy Prize for 1970 and of the Templeton Prize for 1987